Coal Resources of the Americas

Selected Papers

Edited by

Frank E. Kottlowski *New Mexico Bureau of Mines and Mineral Resources, Socorro, New Mexico 87801*
Aureal T. Cross *Michigan State University, Department of Geology, East Lansing, Michigan 48824*
A. A. Meyerhoff *P.O. Box 4602, Tulsa, Oklahoma 74104*

SPECIAL PAPER 179

THE GEOLOGICAL SOCIETY OF AMERICA
3300 Penrose Place, Boulder, Colorado 80301

Library of Congress Catalog Card Number 78-57145
ISBN 0-8137-2179-2

Published by
THE GEOLOGICAL SOCIETY OF AMERICA, INC.
3300 Penrose Place
Boulder, Colorado 80301

Printed in the United States of America

Contents

Introduction

In the early 1500s, the Spanish swarmed over much of South America, Central America, and the southern part of North America in search of gold. They found much yellow gold — some of which had been accumulated through several centuries by the early Americans — and exported much of it back to Spain. Today only a few scattered mines produce yellow gold in Latin America. But more important to modern economies is the black gold of the Americas, the fossil fuels. Much of this is liquid black gold, crude oil, so copiously produced in countries such as Argentina, Colombia, Ecuador, Mexico, Peru, and Venezuela. Unlike the United States and Canada in central and northern North America, Latin America has not been blessed with extensive reserves of the solid black gold, coal. With reserves of crude oil and natural gas being depleted rapidly, the coal and lignite resources of Latin America have become increasingly more vital for the production of energy and chemicals in the countries south of the Rio Grande — or the Río Bravo del Norte as it is called in Mexico.

The last broad appraisal of the coal resources of Latin America was for the 12th International Geological Congress in 1913. Much additional exploration for coal has been completed, and many new fields have been developed during the past 63 years. With the present world energy crisis, it is especially appropriate that this summary of the coal and lignite resources of the Americas, prepared by experts in each country, be made available to aid in economic planning.

In 1966 the Coal Geology Division of the Geological Society of America, in planning a symposium for its part in the GSA meeting that was to be held in Mexico City in 1968, selected the theme "Coal Resources of the Americas." Because extensive literature was already available on the vast coal deposits of the United States and Canada and because the meeting was being held in one of the principal capitals of Latin America, the symposium was to concentrate on the coal resources of Latin American countries.

From the start, communication was the greatest problem. The language differences of Spanish, Portuguese, and English were amplified by the differences in the various types of Spanish spoken and written in the Latin American countries. The crowded schedules of coal-geology experts and delays or losses by postal systems and other breakdowns in communication made it difficult to conclude arrangements with the speakers. Later, these same problems hampered the gathering of the symposium presentations into a volume for publication.

The lack of travel funds for participants from the Latin American countries to attend the GSA meeting in Mexico City also was a problem. The Coal Geology Division of the Geological Society of America had no source of revenue to finance such costs. The Society's budget had only limited travel funds, and the National Science Foundation considered the symposium to be of such a nature that it should be sponsored by commercial funding. We, the symposium organizers, contacted several private foundations with international programs and the Pan American Union for help but to no avail. Thus, an apparent dead end was reached by the middle of 1967. Finally, the International Union of Geological Sciences recognized the scientific and industrial worth of such a symposium and provided $3,250 for travel funds for the participants, for which we are most grateful.

Summary papers were requested from geologists and other coal scientists from all the countries of Latin America that had any type of coal resources. The papers planned for presentation at the symposium in Mexico City were to be on coal resources of Argentina, Brazil, Central America, Chile, Colombia, Ecuador, Mexico, and Peru; on lignite in the United States; and on Latin American coal reserves as a whole. The author from Chile was unable to attend. Arrangements were finally concluded in the last few weeks preceding the meeting for two coal geologists from Brazil, who had not been able to submit abstracts, to come to Mexico City and participate in the symposium. Many of the papers were presented in Spanish; these were followed by a short review of the paper in English by the multiple chairmen of the session, Aureal T. Cross (USA), Salvador Villamar (Mexico), Jesús Ojeda-Rivera (Mexico), and Frank E. Kottlowski (USA).

The plan of the symposium was to have the papers presented in Mexico City and the manuscripts turned in by the authors at that meeting. It was hoped that the final papers could be published in both Spanish and English (or in the case of the Brazilian papers, in both Portuguese and English), but the costs of publication for this volume have limited us to publishing the papers in English only.

Most of the manuscripts were turned in at the Mexico City symposium, but with few exceptions, entire texts were in the Latin American languages. Aureal Cross and Frank Kottlowski undertook the tasks of translation and editing with considerable secretarial aid from the New Mexico Bureau of Mines and Mineral Resources and from the Michigan State University Department of Geology. Drafting of scientific illustrations was done by the New Mexico Bureau of Mines and Mineral Resources. Differences in interpretation of Spanish and English and almost unbelievable postal service delays bogged down our editorial process to a point of despair that the symposium would ever be assembled. Carbon copies of the correspondence, excluding manuscripts, filled a file 4 inches thick.

In 1974 the editorial aid of Arthur A. Meyerhoff was enlisted. With his knowledge of and fluency with scientific Spanish and Portuguese and his widespread contacts in Latin America, Art was the catalyst needed to push the manuscripts over the last but more difficult humps of updating, final editing, and final acceptance by the authors and submission to GSA for publishing.

Many individuals and organizations deserve our appreciation; to the authors in particular we say, "muchísimas gracias por sus obras y su paciencia." We give special thanks to Guillermo P. Salas, Jesús Ojeda-Rivera, Diego A. Córdoba, Salvador Villamar, Edwin R. Landis, Charles L. Pillmore, Bion H. Kent, Ralph L. Miller, R. Trümpy, W. P. van Leckwijck, and the International Union of Geological Sciences for their various contributions to the success of the symposium and to Russell R. Dutcher, Chairman, Coal Geology Division of GSA, for the Mexico City meeting. Intermediate drafting of some of the illustrations was done by William E. Arnold and Robert L. Price, New Mexico Bureau of Mines and Mineral Resources, and much of the final drafting was executed by Kathryn L. Meyerhoff. To our Latin American amigos and to the parts of Latin America where el carbón es rey, we dedicate this volume.

Frank E. Kottlowski

Aureal T. Cross

A. A. Meyerhoff

Geological Society of America
Special Paper 179

Chilean, Argentine, and Bolivian coals

Hector Flores-Williams*
Department of Geology, University of Chile, Santiago, Chile

ABSTRACT

The coal-bearing strata of Argentina, Bolivia, and Chile crop out along the flanks of the Andes Mountains. Carboniferous coals are known in Argentina from the provinces of San Juan, Mendoza, Neuquén, and Chubut. The Middle and Upper Carboniferous coals are anthracite. Permian coals are known from the provinces of Córdoba, San Juan, Mendoza, and Chubut in Argentina; from the vicinity of Lake Titicaca in Bolivia; and from La Ternera and Quilacoya in Chile. The Permian coals also are anthracites. However, numerous faults and folds, a high ash content, and the limited extent of the coal-bearing basins make these Paleozoic coals of little economic interest. The same is true of the Mesozoic coals that are known from Argentina in the provinces of San Juan, Mendoza, and Neuquén.

The only important and commercial coal deposits from this three-country region are Tertiary coals in Argentina and Chile. In Chile the proved and probable bituminous coal resources total about 100 million metric tons, most of which is in the Eocene Arauco basin, which contains the best-quality coal in Chile. The calorific value of the best of this coal is 13,500 Btu/lb. The Oligocene subbituminous coals of the Magallanes basin in Chile are considered to be one of the largest reserves in South America — calculated to be 5,400 million metric tons. This is in addition to the 380 million metric tons of reserve in the Argentine Río Turbio coal deposits. The Magallanes and Río Turbio coals have a high ash content and a calorific value of 8,500 Btu/lb.

INTRODUCTION

The purpose of this paper is to synthesize what is known about the coal geology and coal reserves of Argentina, Bolivia, and Chile. I deliberately omit the description of peat deposits because of their relative unimportance. They are located mainly in the Patagonian parts of Argentina and Chile, as well as in the high ranges of Bolivia.

Detailed studies of coal petrography are scarce. One important study which has been published is that on the Lota and Schwager coals (Reccius, 1942).

In the sections that follow, I have omitted statistical compilations in order to keep the work brief and to the point. It may be said that coal production in Chile and Argentina has been relatively constant and is an important part of the economics of both nations. In contrast, coal production in Bolivia is nil, and only a few peat deposits are mined.

CHILEAN COALS

The first knowledge of coal in Chile dates to the colonial era, but the first commercial development did not take place until the middle of the last century in the Concepción and Arauco Bay areas. Exploitation began in Arauco Province in 1866, in the Magallanes region in 1897, and in Valdivia in 1910 (Figs. 1, 2).

The average coal production from 1941 to 1966 was 1,862,000 metric tons per year. Of the 1,600,000 metric tons (hereafter, t) of coal consumed in Chile during 1966, 95 percent came from Chile. Of this amount, 70 percent (1,070,000 t) was mined at Lota and Schwager (Fig. 3). The remainder (450,000 t) came from Arauco, Valdivia, and Magallanes (Figs. 1, 2).

The Lota and Schwager coals are mined from depths that are below sea level. The mines are 7 km from the coast (Figs. 3, 4).

Coal Classifications

The Chilean coals generally are classified as anthracite, bituminous coal, lignite, and peat. The bituminous coal and lignite are the only ones of economic importance. Coals having a calorific value greater than 11,320 Btu/lb are considered to be bituminous; those having a smaller calorific value are considered to be lignite. The physical and chemical characters are shown in Table 1. According to the ASTM norms (American Society for Testing and Materials), the Arauco coals correspond to bituminous A coal and the Magallanes, Valdivia, and Talcahuano coals correspond to bituminous B and C coals.

The Tertiary coals were formed in paralic basins, whereas the Rhaetic coals appear to have been deposited in basins that may be characterized more properly as "mesoparalic."

*Present address: Minmetal, Casilla 2695, Santiago, Chile.

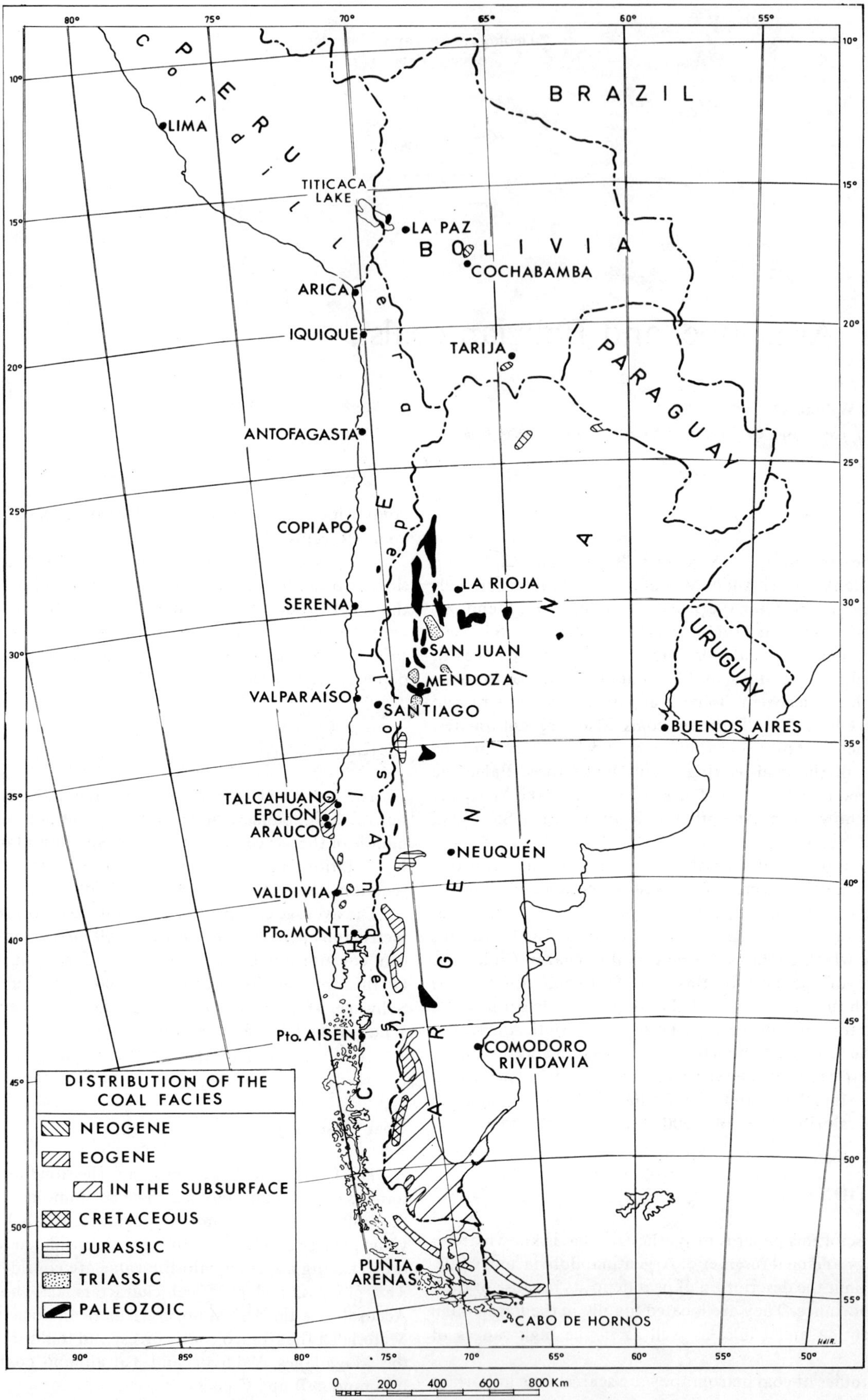

Figure 1. Distribution of Chilean, Argentine, and Bolivian coals by area and age.

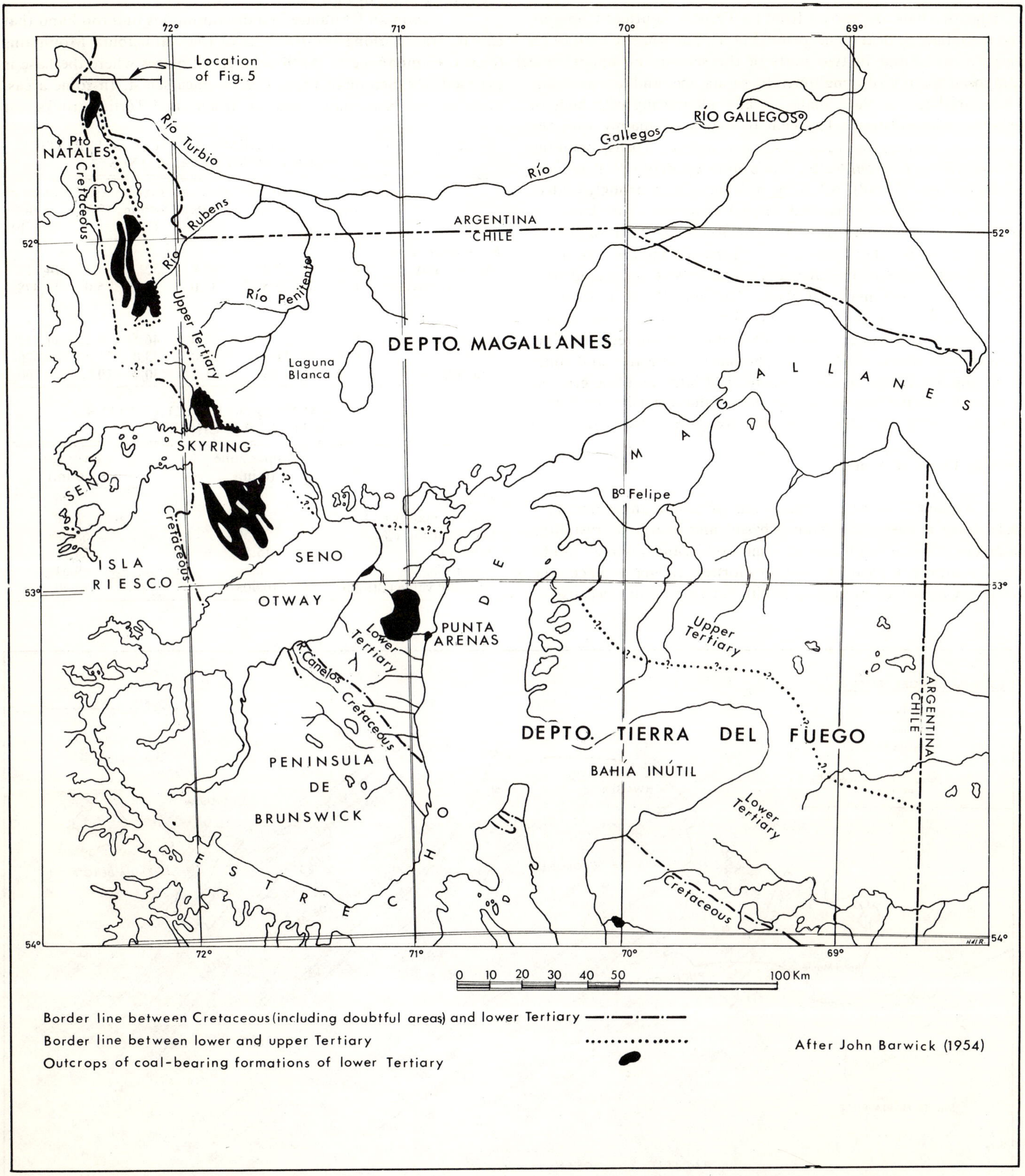

Figure 2. Distribution of coal deposits in the Magallanes–Tierra del Fuego area, southern Chile. From Barwick (1954).

Rhaetic Coals

Rhaetic coals are known from La Ternera, south of Copiapó, and in Quilacoya near Concepción (Fig. 1). The Rhaetic coals of La Ternera are found in two parts of the section: an upper zone composed mainly of conglomeratic sandstone, and a lower zone where argillaceous slaty shale prevails, alternating with beds of fine-grained sandstone. The coal is associated mainly with the sandstone. Various fossil plants occur with the coal beds, including *Dictophyllium carlsoni* Nath., *Clathropteris polyphyllia* Brgt., and others (Brüggen, 1918). All of the beds have been strongly folded. The coals should be classified as coal slates with thin lenses of brilliant coal, accompanied by a large ash content. As shown by the characteristics listed in Table 2, the La Ternera coals are anthracite.

The anthracite of Quilacoya is found in slate and sandstone together with the remains of plants. The flora (Veijl, 1962) is the same as that at La Ternera. The coal-bearing zone consists of a lower continental sequence with anthracite and plant remains. This is overlain by a marine section of alternating lutite and sandstone; at the top is a limnic section composed of lutite and fine-grained sandstone bearing the remains of plants and a freshwater fauna. The La Ternera and Quilacoya coals are noncommercial.

Tertiary Coals of Arauco

In Arauco Province, Tertiary and/or Senonian sedimentary rocks form a basin (the Arauco basin) and overlie a crystalline basement of metamorphic rocks — mica schist and phyllite — and a massive diorite typical of Andean diorites. In many places the Tertiary-Cretaceous boundary is difficult to establish because of the absence of basal conglomerate. The beds are completely intergradational in most areas.

The Senonian strata are in a discontinuous outcrop band that can be traced along the foothills of the Nahuelbuta Mountains from Carampangue to Antihuala. In all places where the base is exposed, the Senonian rocks overlie mica schist. In some areas, however, the Senonian rocks are absent, and Tertiary rocks rest

TABLE 1. PHYSICAL AND CHEMICAL CHARACTERISTICS OF CHILEAN COMMERCIAL COALS

	Specific gravity	Moisture (%)	Volatile matter (%)	Fixed carbon (%)	Ash (%)	Calorific value (Btu/lb)
Bituminous Coal						
Arauco Bay	1.28	2.8	41.8	52	3.4	14,017
Arauco Province	1.27	3.5	37.0	54	5.0	13,478
Lignites						
Talcahuano Bay	1.26	12.4	35.0	46.5	6.1	11,232
Valdivia	1.40	11.7	35.0	33.0	21.0	8,626
Magallanes	1.36	17.6	41.7	30.7	10.0	8,560

TABLE 2. ANALYSES OF RHAETIC COALS, LA TERNERA, CHILE

	Upper zone (6 in.)	Lower zone (20 in.)
Moisture	5.00	4.20
Volatile matter (%)	5.00	5.10
Fixed carbon (%)	74.50	65.95
Ash (%)	15.50	24.75
Sulfur (%)	0.54	9.51
Calorific value (Btu/lb)	10,908	9,650

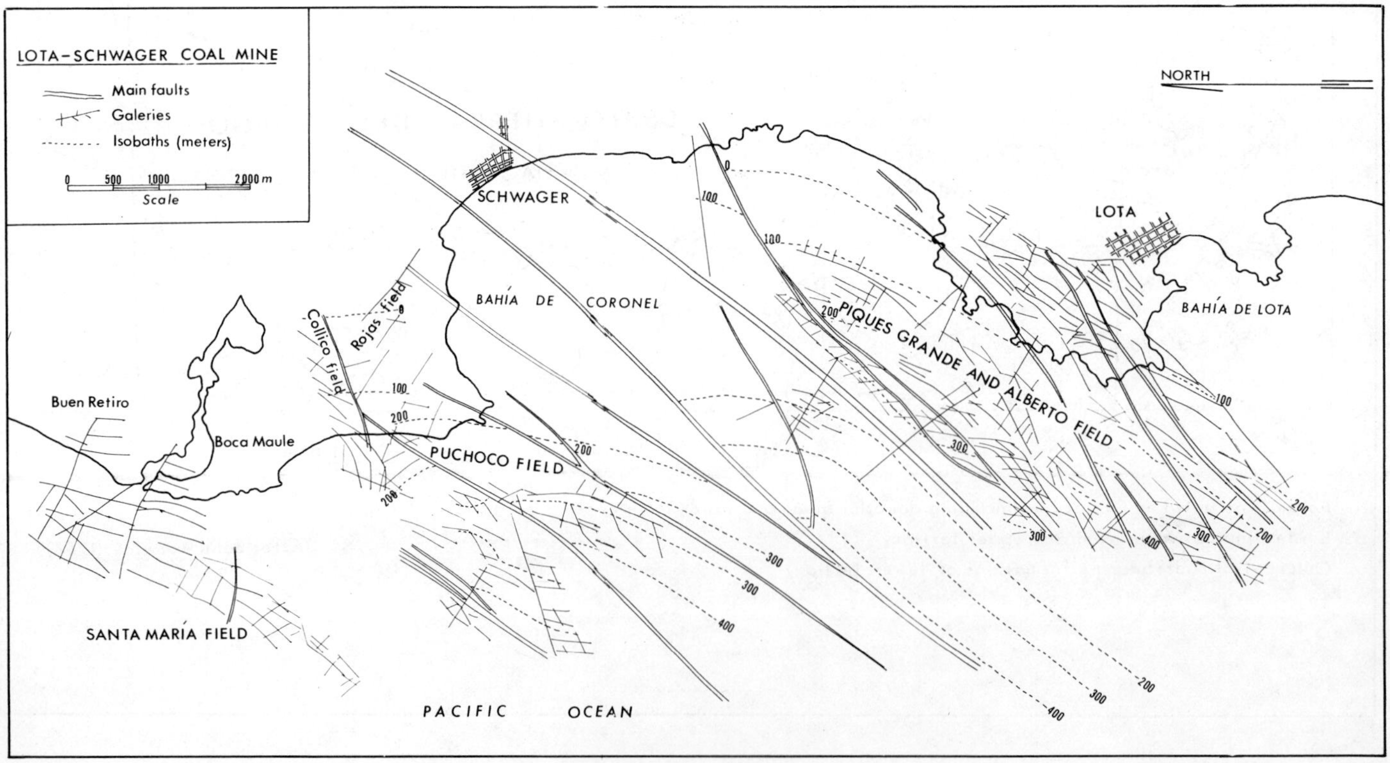

Figure 3. Lota-Schwager mines area, Arauco basin, Chile.

directly on the mica schist of the basement (for example, in Curanilahue).

The base of the Senonian section is a thick conglomerate containing boulders of mica schist, muscovite gneiss, biotite gneiss, and quartzite (with siliceous cement). The general characteristics of the basal Senonian rocks suggest that this conglomeratic zone formed along a coast line.

Above the conglomerate zone are beds of sandstone that become finer grained upward. No more conglomerate beds appear in the sequence. The sandstones are formed of quartz, feldspar, biotite, and muscovite. The matrix is argillaceous and ferruginous. The feldspar and biotite appear to be strongly altered by weathering. The general interpretation is that they were deposited in an intermediate to mild climate.

The Tertiary rocks of Arauco (Fig. 4) are summarized in Table 3 (Muñoz, 1946). Rocks of the Curanilahue stage consist of alternat-

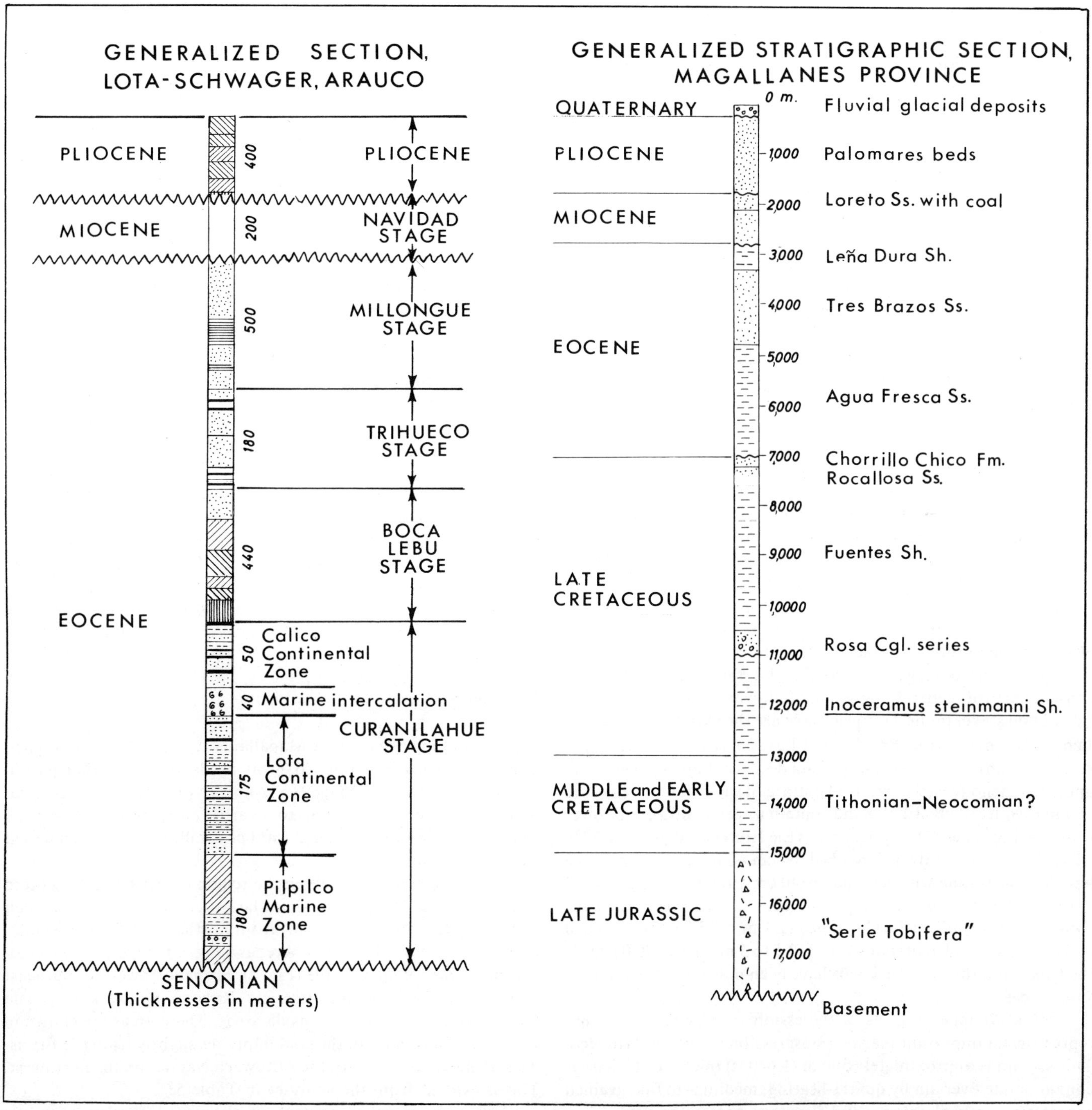

Figure 4. Columnar sections of Lota-Schwager mines area, Arauco basin (left), and of Magallanes area (right), Chile.

ing marine and continental coal-bearing sedimentary strata. The beds of the Boca Lebu stage are entirely marine. In contrast, the Trihueco stage consists of alternating continental coal-bearing strata and marine strata; the continental, coal-bearing strata predominate. The Millongue stage consists entirely of marine rocks. The Miocene Navidad stage in the Arauco region also is marine.

The Curanilahue stage is subdivided from top to base into (1) the Colico continental zone, (2) the marine intercalation, (3) the Lota continental zone, and (4) the Pilpilco marine zone (Fig. 4).

The Pilpilco marine zone formed during a transgression that covered a large continental platform area. Sandstones are widely distributed on this former platform. The thickness at Colico of the Pilpilco zone is 180 m (590 ft). The overlying Lota continental zone is mainly continental. At the base is a thick sandstone overlain by finer-grained sandstone interbedded with clay and coal. The total thickness of this zone is 175 m at Lota (574 ft). At the Lota mine, the Lota zone is subdivided as shown in Table 4. Eastward, the three coal beds are separated by layers of sandstone and claystone. The "marine intercalation" zone is 42 m thick (138 ft) at Lota and 56 m thick (184 ft) at Schwager. It consists of green, fine-grained sandstone beds, at the base of which a few conglomerate layers locally are present. The Colico continental zone is well developed at Lota, but has very few coal beds. The coal beds assume great importance in the Peumo-Maquehua area (45 km south-southwest of Concepción; Fig. 1) where two beds are mined (at Peumo)—the Colico N bed and the Colico S bed. The Colico zone ranges in thickness from 30 to 100 m (98 to 328 ft).

The strata of the Boca Lebu marine stage were formed during a large-scale transgression onto the Curanilahue beds. The upper part (50 m = 164 ft) consists of coarse- to fine-grained argillaceous sandstone interbedded with thin layers of claystone containing some plants. Minor conglomerate beds are present in the upper part of the central part. The entire central part is fossiliferous. The lower part of the Boca Lebu stage (46 m = 151 ft) consists of ashy claystone with thin intercalations of plant-bearing clay, some of them coalified. The thickness of the lower part is 550 m (1,804 ft). Although the section contains few fossils, the widespread presence of glauconite in this lower part suggests that the section is marine.

The Trihueco stage is a continental series deposited as a consequence of general regression following deposition of the Boca Lebu stage. In the Trihueco stage are the various coal beds of Lebu, Millongue, Huenapidén, and Lavapié (western part of the Arauco basin), and of Trihueco, Pata de Vaca, Lajuela, Chupalla, and El Huacho (eastern part). The strata of the Trihueco stage are 180 m (590 ft) thick at Lebu and contain upper, central, and lower coal-bearing zones. The upper zone is the Huiters bed (0.6 to 1.37 m = 2 to 4.5 ft) and the Broncienta bed (60 cm = 2 ft). The central zone has three beds: one without a name (40 cm = 1.3 ft), the Ratón bed (15 cm = 0.5 ft), and the Chiflón (76 cm = 2.5 ft). The lower zone consists of three beds: the Chico bed (1.1 m = 3.5 ft), the Alto bed (1.52 m = 5 ft), and an unnamed bed (0.5 to 3 m = 1.6 to 10 ft). Only the Chico and the unnamed beds have been mined economically in some areas.

The Millongue stage is a succession of marine strata that represents an important marine transgression above the Trihueco. This section is approximately 500 m (1,640 ft) thick. At the base is conglomerate overlain by quartz-bearing, medium- to fine-grained sandstone, sandy claystone, and claystone containing peat.

TABLE 3. SUBDIVISION OF TERTIARY SECTION OF ARAUCO, CHILE

Age		Unit
Pliocene		Present surface
	Angular unconformity	
Miocene		Navidad stage
	Angular unconformity	
Eocene		Millongue stage
		Trihueco stage
		Boca Lebu stage
		Curanilahue stage

TABLE 4. SECTION AT LOTA MINE, ARAUCO

Seam	Correlation with Schwager mine	Thickness (m, ft)	Sandstone and clay between seams (m, ft)
Upper (Arriba)	Schwager 3	1–1.5 (3.3–4.9)	37 (121)
High (Alta)	Schwager 5	1.6 (5.2)	9 (30)
Small (Chica)	Schwager 4	0.8 (3.0)	

The Navidad stage is marine and does not have any intercalated continental strata.

The Pliocene strata of Arauco Province crop out extensively. The Pliocene units are petrographically uniform, consisting of sandy claystone and argillaceous sandstone that is weakly cemented and contains abundant fossils. The thickness is approximately 400 m (1,312 ft).

Tectonically, the total section is characterized by major unconformities at the base of the Senonian section, between the Senonian and the Eocene sections, between the Eocene and Miocene sections, and between the Miocene and Pliocene sections. The major structures in the area are normal faults. In general they are parallel with one another. They strike north-northeast and northeast in Lota-Schwager (Fig. 3), Colico, Millongue, and Millaneo; due north at Curanilahue; and north-northwest to north-northeast at Lebu, south of the area of Figure 3. At Lebu there also is an east-striking set of faults. "Hinge faults" are found throughout the area.

Magallanes Coals

Coal mining began in the Magallanes region in 1897 (see Fig. 2). The original mining supplied local needs and ships that passed through the Strait of Magellan. Another purpose of the original mining was to supply coal to nearby areas in the Argentine Republic. It is estimated that a total of 5 million t have been mined from this region.

The most important coal in the region comes from the Loreto Formation (the Oligocene part). The Loreto unit extends in a belt approximately 200 km (125 mi) from northeast of Puerto Natales to Porvenir, close to Punta Arenas (Fig. 2). Five distinct coal-bearing areas are identifiable (Fig. 2). From northwest to southeast these are termed the Natales, Río Rubens, Seno Skyring, Isla Riesco, and Brunswick Peninsula areas. These areas are shown in Figure 2. The quality of the coal improves slightly from the Brunswick Peninsula on the southeast toward Natales on the northwest. This is evident from the analyses in Table 5.

The high moisture content of the coal, together with the

TABLE 5. CHARACTERISTICS OF MAGALLANES COALS

Area	Mine	Specific gravity	Moisture (%)	Volatile matter (%)	Fixed carbon (%)	Ash (%)	Sulfur (%)	Calorific value (%)
Brunswick Peninsula	Loreto	1.27	25.25	30.25	31.90	12.60	1.50	7,560
	Vulcano	1.20	23.60	30.40	32.45	13.55	0.30	7,100
	Tres Puentes	1.21	27.20	32.00	30.90	9.40	0.45	7,460
	Rocu	1.41	23.00	28.65	34.30	14.05	0.50	6,960
	El Chino	1.35	22.80	31.95	28.55	16.70	3.05	6,940
	Servidora	1.42	18.90	31.80	26.15	23.15	3.65	7,400
	Pecket	1.29	18.50	34.00	40.50	7.00	0.45	8,730
	Average	1.31	23.00	31.25	32.20	13.55	1.35	7,480
Isla Riesco	Elena	1.31	17.35	33.00	37.80	11.85	0.40	8,810
	Tres Hermanos	1.29	18.50	34.05	40.50	6.95	0.40	8,750
	Josefina	..	15.05	36.25	36.90	11.80	0.45	9,940
	Average	1.30	17.00	34.40	38.40	10.20	0.40	9,080
Natales	Chilena	1.30	8.50	33.30	40.30	17.85	0.50	10,320

argillaceous character of the coal ash, makes the coal extremely brittle, which is a major adverse factor in developing the coal commercially for fuel.

Much of the information on these coals is contained in unpublished reports of the ENAP (Empresa Nacional de Petróleo), particularly reports by Barwick (1954), Wenzel (1955), Cortés (1966), and González (1966). In these reports there are several evaluations of the coal possibilities in the Magallanes area. These are summarized below.

Brunswick Peninsula. The coal in this region crops out in the area of the Río de las Minas, in a broad monoclinal structure dipping toward the northeast. It occupies an area of about 100 km^2 northwest of the city of Punta Arenas. Within this coal-bearing area is the Estancia Mina Rica.

Within the coal-bearing area, there are five coal beds with a total thickness of 9 m (30 ft) that are present in at least two-thirds of the outcrop and subcrop area. At Mina Pecket, along the eastern coast of Seno Otway, one coal bed has a thickness of 3.6 to 5.5 m (12 to 18 ft). The coal occurrences at Mina Pecket and in the Punta Arenas area are at the margins of a large synclinal zone of approximately 50 km^2 that extends between Mina Pecket on the northwest and Estancia Mina Rica on the southeast. The synclinal area between these two coal-bearing zones is overlain by a thin veneer of glacial deposits. It is quite probable that the thick coal bed that is found at Mina Pecket subcrops beneath this thin veneer and that it contains an important reserve.

Isla Riesco. Although there are several coal-bearing areas, three strongly folded synclinal zones in the northern part of the island are of particular interest. These are Mina Elena, Río Patos, and Chorrillo Watt. The coal exhibits marked lateral facies changes. For example, at Mina Elena, the coal is at the base of the Loreto and has a thickness of 6 m (20 ft). In the southern part of the same syncline at the same stratigraphic level, in the upper part of the Río Eduardo, very close to Seno Otway (Fig. 2), there are three coal beds instead of one. The three beds have a combined thickness of 5.1 m (17 ft). In addition, there are less important coal beds within this area at different stratigraphic levels.

In the Río Patos syncline, between the Chilenita and Josefina mines, there are 8 to 10 coal beds cropping out near the coast. The thicknesses of these seams range from 1 to 2 m (3.3 to 6.6 ft). The combined thickness of the 8 to 10 seams is about 15 m (50 ft).

Farther toward the southeast—approximately 18 to 20 km—studies have been made in the upper part of the Río Aracielis drainage basin. These studies established the presence of two coal seams at the base of the Loreto with thicknesses of approximately 2 to 12 m (6.6 to 39 ft).

On the same peninsula, in the Chorrillo Watt syncline (which extends for about 10 km to the upper part of the Río Vaquería area), there are two coal beds (which are covered abruptly by a glacial veneer). These two beds have a total thickness of 3.6 m (12 ft). They are unknown along the coast. In the Río Vaquería area, there also are two coal beds with a total thickness of 9 m (30 ft). One of these beds reaches a thickness of 7 m (23 ft).

Seno Otway — North Zone. This zone is on the northern side of Seno Otway and south of Seno Skyring (Fig. 2). It attracted very little attention because of the small thicknesses and poor quality of the beds that had been worked at Mina Marta and at Punta Carbón, close to the coast of Seno Otway. However, in 1950, along the lower course of the Río Pérez, a bed 2 m (6.6 ft) thick was discovered. This bed, combined with the other two beds, brings the total thickness of coal in this region to 3.5 m (11.5 ft). Furthermore, the presence of these beds can be projected through an area of approximately 40 km^2. These beds occur in anticlines and synclines.

Río Rubens. This is the area along the north side of Seno Skyring (Fig. 2). It is a large zone of rather open folds centered around Río Rubens. Unfortunately, knowledge of this area is rather scanty. One bed 6 m (20 ft) thick and two beds totaling 1.25 m (4.1 ft) are known in Chorrillo Carbón, as well as in the upper part of the Río Rubens and the Río Negro.

Natales. Lack of systematic mapping data does not permit at this time a good evaluation of this area. However, the beds present here are known to be continuous from the Río Turbio deposits of Argentina. To date, only one bed with a thickness of 1.3 m (4 ft) is known—in Mina Chilena. This mine is very close to the Argentine frontier.

Although less than 10 percent of the coal resources of Magallanes can be considered as "proved," the geologic data suggest that minimum resources are 5 million t. The coal resources are discussed in the following section.

TABLE 6. CHILEAN BITUMINOUS COAL RESOURCES (MAINLY ARAUCO AREA)

Area	Proved ($\times 10^3$ t)	Probable ($\times 10^3$ t)	Possible ($\times 10^3$ t)
Lota-Schwager*	3,000	40,000	38,000
Lota-Schwager†	..	..	140,000
CAP Concession, Arauco§	..	..	150,000
CORFO Concession, Trihueco	2,000	3,000	30,000
Colico	10,000	40,000	50,000
Total	15,000	83,000	408,000

*Less than 1,000-m depth.
†More than 1,000-m depth.
§Unpublished report, Toenges Mission.

CHILEAN COAL RESOURCES

Bituminous Coals

The resource figures of bituminous coals are given in Table 6. The information was taken from unpublished reports of the Compañía Lota Schwager, the Corporación de Fomento de la Producción, the Servicio de Minas del Estado, and the Instituto de Ingenieros de Minas. The proved and probable resources, as shown in Table 6, total about 100 million t with a calorific value of 13,500 Btu/lb. Within the Arauco basin, an area of only 120 km^2 was taken into account. The total area of the basin is 2,000 km^2, a fact which indicates that future investigations may increase the resource considerably.

TABLE 7. CHILEAN SUBBITUMINOUS COAL RESOURCES

Area	Subarea	Proved ($\times 10^3$ t)	Probable ($\times 10^3$ t)	Area of possible reserves* (km^2)
Bahía de Talchahuano		1,500	1,500	2,000 (submarine)
Valdivia and Llanquihue		2,000	3,000	
Subtotal		3,500	4,500	2,000
Magallanes	Natales	1,500	13,500	15,000
	Río Rubens	50,000	450,000	
	Seno Skyring	55,000	495,000	
	Isla Riesco	322,000	2,898,000	
	Brunswick	111,500	1,003,500	
Subtotal		540,000	4,860,000	15,000
Total		543,500	4,864,500	17,000

*Basin surface areas.

TABLE 8. CHILEAN COAL PRODUCTION, 1960-1974

Year	Amount ($\times 10^3$ t)	Year	Amount ($\times 10^3$ t)
1960	1,471	1967	1,476
1961	1,764	1968	1,611
1962	1,855	1969	1,704
1963	1,719	1970	1,510
1964	1,789	1971	1,623
1965	1,724	1972	1,427
1966	1,652	1973	1,390
		1974	1,520

TABLE 9. CHILEAN COAL PRODUCTION, 1970–1974

Zone	1970	1971	1972	1973	1974
			($\times 10^3$ t)		
Concepción*	1,122	1,230	1,002	997	1,058
Arauco*	325	328	362	336	421
Subtotal	1,447	1,558	1,364	1,333	1,479
Valdivia†	29	30	33	32	28
Magallanes†	34	35	30	25	13
Subtotal	63	65	63	57	41
Total	1,510	1,623	1,427	1,390	1,520

*Bituminous.
†Subbituminous.

Subbituminous Coals

These resources are shown in Table 7 and correspond to the subbituminous B and C coal categories of the ASTM specifications. It can be seen from Table 7 that the proved and probable resources of the Magallanes area alone total 5,400 million t and therefore constitute one of the greatest coal resources of South America. The calorific value is estimated to be 8,500 Btu/lb on the average.

Production and Miscellaneous Information

During 1975, the Corporación de Fomento de la Producción (a Federal organization), began to restudy the Magallanes coal resources by drilling test holes. The purpose of the drilling was to find sufficient coal for shallow open-pit mining. The results of the drilling showed that in the southern part of the Arauco basin, at approximately lat 41°S, there are new resources of coal. However, the data are too scanty to draw any firm conclusions at this time.

A study has been begun to determine what Chilean coals can be used for coking purposes.

Production since 1960 is shown in Table 8. Production during the period 1970–1974 is shown in Table 9.

ARGENTINE COALS

From the very beginning of Argentina's history, the government has officially promoted coal mining. However, important coal exploration began only during the beginning of this century; this exploration was in Patagonia. Two important works on the development of Argentina's coal industry are by Angelelli (1941) and Borello (1956).

Carboniferous Coals

Coals formed during the Carboniferous Period are found principally in the provinces of La Rioja, San Juan, Mendoza, and in the northern and south-central parts of Patagonia. The coals are interbedded and associated with limnic and mesoparalic facies. This is evident from a study of the various facies. The distribution of the coals of this age is shown in Figure 1.

Within the continental facies of the Carboniferous coals, there are many floral associations such as the *Rhacopteris ovata* Walk., commonly found in coal beds that are exploitable. A particular notable *Rhacopteris ovata* coal is the La Negra deposit in the Carboniferous coal zone of La Rioja. This plant association seems to have appeared at the end of Early Carboniferous time and developed most prominently during Middle and Late Carboniferous time. The basins in which these coals developed were either very small in extent or the area in which coal deposition took place was rather small. Therefore, such basins are of relatively small importance. Limnic facies are extensive west of Catamarca, in the middle and western parts of La Rioja Province, and in the northern and central parts of Mendoza Province (Fig. 1). In contrast, the mesoparalic deposits are found particularly in La Rioja and San Juan Provinces.

Study of these deposits indicates that local tectonic conditions were such that the coal seams are rather thin and contain many laterally extensive clay-shale breaks. The thickness of the seams generally is 30 to 60 cm (1 to 2 ft). For this reason, the deposits are noncommercial, and the resources are small. Characteristics of these coals are shown in Table 10.

Rhaetic Coals

The Rhaetic (Triassic) strata of Argentina are developed in continental facies. The deposits have a limnic character. Coal began to form in small basins. The coal beds are irregular and discontinuous. Only briefly during the middle part of Rhaetic time was there important coal formation in San Juan and Mendoza Provinces. The deposits of Misiones, San Luis, Neuquén, and Santa Cruz Provinces are much smaller and are of minor importance.

TABLE 10. CHARACTERISTICS OF CARBONIFEROUS COALS, ARGENTINA

Characteristics	Mina Cortadera, La Rioja Province	Mina Delfina, San Juan Province
Moisture (%)	2.19	3.27
Volatile matter (%)	30.94	23.37
Fixed carbon (%)	53.88	45.18
Ash (%)	12.99	28.18
Sulfur (%)	1.45	1.48
Btu/lb	12,700	9,800

TABLE 11. RHAETIC COAL OF THE MARAYES DEPOSIT, SAN JUAN PROVINCE, ARGENTINA

Characteristics	Value
Moisture (%)	2.6
Volatile matter (%)	24.1
Fixed carbon (%)	41.5
Ash (%)	31.8
Sulfur (%)	5.56
Btu/lb	9,150

One of the best-known coal deposits is that at Marayes, San Juan Province (Fig. 1). The coal-bearing section has six seams of coal that change thickness from one place to the other. The lowest layers (I and II) are separated by 17 to 23 m (55 to 76 ft) from the upper layers. The upper layers characteristically are separated from one another by clay-mudstone and other terrigenous materials in zones 6 to 7.5 m (21 to 25 ft) thick. Layers I and II are the thickest, being 28 to 51 cm (11 to 21 in.) thick. In many places they are broken up by clay layers, so that the thickest solid unit of coal is only 18 cm (7 in.) thick. The beds dip from 6° to 10°.

There has been some exploitation of this coal during the past century, and four boreholes were drilled by the Argentine Dirección de Minas. On the basis of these explorations, it was determined that in an area of 10 km^2 (4 mi^2), there is a coal resource of 2.5 million t. The average thickness of each bed was assumed to be 30 cm (1 ft). The quality of the coal is shown on Table 11. I should note that the fact that the coal seams are extremely thin makes their commercial exploitation rather difficult.

Jurassic Coals

These coals occur in three separate zones within the Jurassic section. The lower coal is in middle to upper Liassic (Lower Jurassic) rocks; the second is in upper Dogger (Middle Jurassic) rocks; and the highest is in uppermost Malm (Tithonian or uppermost Jurassic) rocks.

The famous coals of the Río Atuel region, Mendoza Province (Fig. 1), with the *Diperidecae* flora, belong to the Liassic group of coals. These coals were formed in the continental interior; thus the sea rarely reached the basin of deposition.

The Dogger coals are known mainly from Neuquén, where the coal seam at Charahuilla (Mina Burgos) is within Callovian sedimentary rocks. The Callovian rocks contain remains of *Ctenosreon neuquense* Weax., *Perna americana* Forb., and others.

The Tithonian coals are known particularly from the Cerro Meseta region of Lago San Martín, Patagonia (Santa Cruz Province). They contain fossil *Belemnopsis* sp. that are definitely of Tithonian age.

All of the Jurassic coal accumulations are of limited extent and belong to a paralic facies. In general, the maximum thickness of the seams is 51 to 76 cm (20 to 30 in.). Many of the coals have a large volume of ash, and the resources are limited. Coal analyses from several of the mines (Table 12) give an idea of the quality of these coals.

Cretaceous Coals

The oldest Cretaceous coal deposits known in Argentina are found in the upper part of the Neocomian section at Lago

TABLE 12. CHARACTERISTICS OF ARGENTINE JURASSIC COALS

Characteristics	Arroyo de la Manga mine, Mendoza Province (Liassic)	Arroyo del Pedrero mine, Río Atuel, Mendoza Province (Liassic)	Coihueco mine, Río Atuel, Mendoza Province (Liassic)	Mina Cristina, Neuquén Province (Dogger)
Moisture (%)	0.19	7.07	3.4	8.50
Volatile matter (%)	8.09	5.43	5.5	26.55
Fixed carbon (%)	65.81	78.18	40.6	31.24
Ash (%)	25.91	9.32	50.5	33.71
Sulfur (%)	3.00	0.37	..	..
Btu/lb	11,500	11,000	..	7,600

TABLE 13. TERTIARY COALS, RIO TURBIO COAL BASIN, PATAGONIA

Province	Locality	Facies	Moisture (%)	Volatile matter (%)	Fixed carbon (%)	Ash (%)	Sulfur (%)	Calorific value (Btu/lb)
Neuquén (southeastern part)	Mina Jorge Newbury	Limnic	6.57	27.62	44.13	21.68	N.A.	12,000
Río Negro	Pico Quemado	Limnic	2.50	24.55	47.79	25.38	0.46	11,700
Chubut	Indio	Limnic	10.87	24.89	30.90	33.34	..	7,200
Santa Cruz	Cabo Curioso	Mesoparalic	15.16	23.21	31.49	30.14	..	Very low
Santa Cruz	Arroyo de Carbón, Arroyo de la Cumbre	Mesoparalic	9.18	19.28	29.78	41.76	..	7,190
Santa Cruz	Río Turbio, lower bed	Paralic	9.74	27.94	39.38	22.94	0.57	8,086
Santa Cruz	Río Turbio, lower bed	Paralic	6.6	34.0	44.0	15.4	0.35	12,100
Santa Cruz	Río Turbio, Sierra Dorotea (Dorotea) bed	Paralic	7.0	35.0	45.0	11.0	1.10	12,600

Puyrredón and Río Meyer, western Santa Cruz Province (Fig. 1). The coals are below marine strata that bear remains of *Hatchericeras* sp. cf. H. *terdense* Stant. of Hauterivian age.

In the Comodora Rivadavia region, the coal deposits are very restricted, are distributed irregularly, and contain remains of *Ptilophyllum* cf. P. *acutifolium* Oldh. These beds are of Senonian age.

In the youngest Cretaceous beds of the Patagonian region, the Piedra Parada seam, Chubut Province, is in a sequence of strata that (above the coal) includes remains of *Ostrea rionegrensis* Ih. and *Trigonia windhauseniana* Wilck. These appear to be of Danian age and may be slightly older than or approximately equivalent to the coal seams in the Neocretaceous strata of the Río Turbio region, south of San José.

Thus, all of the Cretaceous coal deposits are restricted to the general territory known as Patagonia. The principal and most important outcrops are in the Piedra Parada region of Chubut and in the upper Río Meyer region of western Santa Cruz Province. All of the Lower Cretaceous coals are believed to have been deposited in a paralic environment.

The possibilities for development of the commercial coals are limited. The beds are very thin and their lateral—that is, areal—extents, as far as is known, are restricted.

Tertiary Coals

The largest coal resources of Argentina are in the Tertiary basins of Patagonia, particularly in the Río Turbio basin that extends westward into Chile. They may be classified as Paleogene. They are easily identifiable by their fossil content as well as by their good degree of development. Marine intercalations date the beds well.

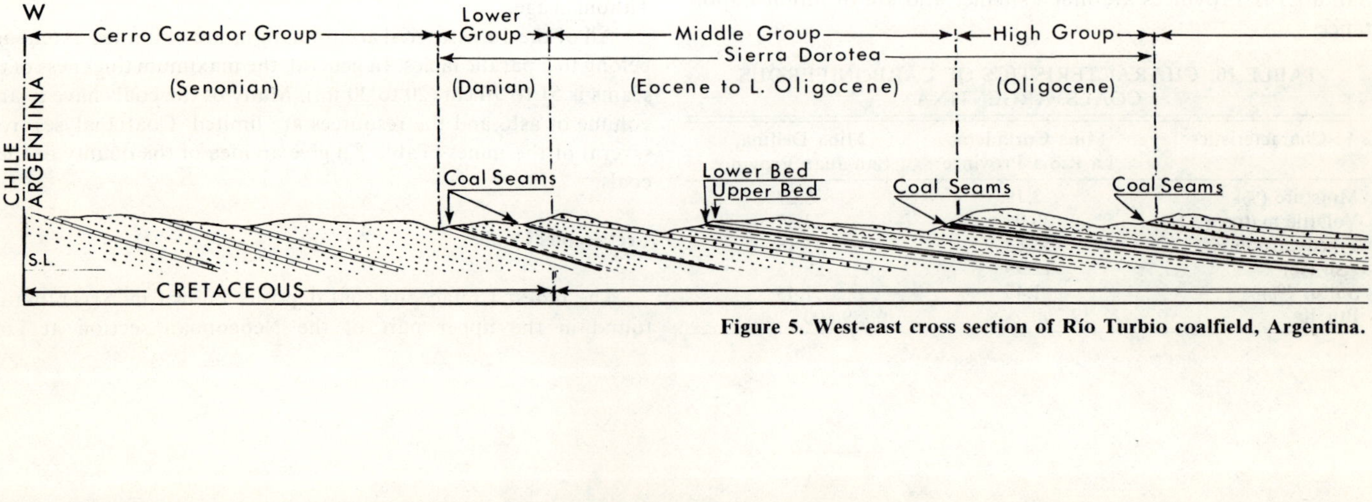

Figure 5. West-east cross section of Río Turbio coalfield, Argentina.

TABLE 14. ARGENTINE COAL RESOURCES

Mine	Age	Proved (×10³ t)	Probable (×10³ t)	Possible (×10³ t)	Total (×10³ t)	Classification
Río Tambillo	..	390	20	20	430	Low bituminous
Negra	..	..	35	..	35	Bituminous
Delfina	Carboniferous	..	..	50	50	Bituminous
Richard CK	..	..	..	200	200	Bituminous
El Saltito	..	..	..	200	200	Bituminous
La Manga	Liassic	..	..	100	100	Anthracite
Burgos	Dogger	..	..	25	25	Bituminous
J. Newbery	Tertiary	..	..	5	5	Bituminous
Pico Quemado	Tertiary	75	650	1,915	2,640	Bituminous
Indio	Tertiary	50	100	750	900	Bituminous
Santa Ana	..	..	..	10	10	Bituminous A
Lepa	..	255	..	..	255	Bituminous A
La Criolla	..	..	20	50	70	Bituminous A
Cabo Curioso	Tertiary	2.5	1.5	1	5	Bituminous
Río Turbio	Tertiary	273,000	74,500	23,000	370,500	Subbituminous
	Total	273,722.5	75,326.5	26,326	375,425	

Note: From Borrello (1956). See Bergmann (this volume) for more up-to-date evaluations of certain areas.

The lower part of the sequence is in the middle part of the Sierra Dorotea Group (433 m = 1,420 ft). At the base of the coal-bearing sequence is *Venericardia* cf. *V. planicosta.* This sequence of coal-bearing strata is concordant on the uppermost Cretaceous strata, which include *Ostrea rionegrensis* Ih. of the lower part of the Sierra Dorotea Group.

The middle part of the Sierra Dorotea Group can be assigned, on the basis of its fossil content, to the Eocene and the lower part of the Oligocene Series. It contains two coal-bearing complexes. Marine and continental layers are interbedded. The highest marine zone contains *Ostrea maxima* Lupé of middle Oligocene age. Unconformably above it is the Latorre Group, 450 m (1,500 ft) thick. The Latorre is equivalent in part to the unit which, in Magallanes, Chile, belongs to the Palomares Formation (late Miocene to Pliocene age).

The Eocene section of the Río Turbio region extends southward into Magallanes, Chile, where it is included with the Boquerón Group. It may also be correlated with the lower part of the sequence of strata at Concepción, Chile (Fig. 1), which contains the faunal association characterized by *Venericardia planicosta* Lamarck.

Paleogeographic study reveals three areas of the Río Turbio coalfield where the coal is either limnic, mesoparalic, or paralic.

The limnic facies is in the northwestern region of Patagonia and includes southern Neuquén, Río Negro, and western Chubut. The mesoparalic beds occur in an extensive area of eastern and western Santa Cruz Province. The paralic facies, on the other hand, are very restricted. The most important paralic zone is in the Río Turbio coalfield, where paralic and marine facies alternate.

The quality of coal is shown in Table 13. The area is large, and the quality is sufficiently good that the coal basin of Río Turbio is the most important one in Argentina today. In the Río Turbio sequence there are at least six coal complexes and productive zones. The lowest belongs to the lower part of the Sierra Dorotea Group. Two other coal-bearing zones are in the middle part of the Sierra Dorotea Group. The upper two are near the top of the upper part of the Sierra Dorotea Group and at the very top of the Río Turbio sequence as a whole (Fig. 5).

The coal within the lower zone has an average thickness of 1.5 m (5 ft) and a maximum thickness of 7.6 m (25 ft). Up to 40 percent of these zones may consist entirely of non–coal-bearing intercalations.

The higher exploitable seam has a thickness that ranges from 1.2 to 1.5 m (4 to 5 ft). It has been exploited on a small scale in the Puerto Natales area of Chile (Fig. 2).

There is a noncommercial bed B at the base of the higher sequence. Above it is bed A, 1 m (3.3 ft) thick, which is exploitable.

The most important part of the entire coal series is in the Sierra Dorotea Group, 5 to 7.6 m (16 to 25 ft) above bed A. The average thickness of the seam is 1.5 m (5 ft). Table 14 shows the coal resources, as calculated by Borello (1956). His figures are very conservative and should be regarded as a minimum.

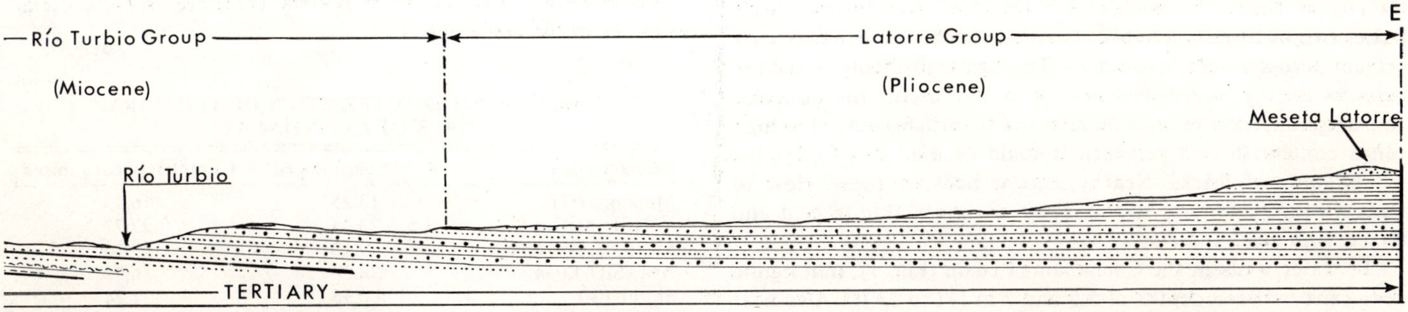

Approximate location is shown on Figure 2. From Borrello (1956).

BOLIVIAN COALS

The works by Ahlfeld (1954) and unpublished reports provided to me by the Corporación Minera de Bolivia form the basis for this section.

The existence of bituminous coal has been known for more than 100 years on the Copacabana Peninsula of the Lake Titicaca region, on the border with Peru. It was first mentioned by Forbes (1861), and by 1875, Agassiz (1875-1876) found the Yampupata mine in full production. The natives of the Siripaca and Yampupata Ranches, adjacent to Lake Titicaca, used the coal to fire lime.

Anthracite

According to various investigations by Derains, Koslowski, and Ahlfeld (see Ahlfeld, 1954), the anthracite is not economically exploitable. The best-known anthracite bed is in the Copacabana Peninsula region and on the Isla del Sol, Lake Titicaca. There it occurs in a syncline in which Permian rocks overlie Devonian rocks with paraconformity. Near the base of the Permian strata, between yellow arkosic sandstone beds, is a thin zone that consists of coal-bearing black slate. The average thickness is 10 m (33 ft). Within the coal-slate are local lenses of anthracite with a few centimeters of thickness. The coal contains 30 to 50 percent of ash and 0.48 to 1.3 percent of sulfur.

Farther south on the Apilapampa Ranch of Cabinota Province, Department of Cochabamba (Fig. 1), where the Permian strata reach a maximum thickness, there are numerous thick beds of marine limestone at the top of the Permian. Just below these thick limestone beds is a zone containing carbonized tree trunks and one thin bed of anthracitic coal. This very thin bed is known from only a single locality.

Lignites

Lignite beds generally are thin in Bolivia. They are known in several of the Pliocene-Pleistocene basins of the Andes. In general, they are horizontal and not deformed. Ahlfeld's (1954) studies of these beds have generally been negative.

The lignite is very immature, earthy, and mixed with clay and sand. Its calorific value (dry) ranges as high as 5,380 Btu/lb. Thus, it is of extremely poor quality.

The most important of the Pliocene-Pleistocene seams was observed by Ahlfeld (1954) in the Tarija basin (Fig. 1). Here are lignite beds ranging in thickness 30 to 90 cm (1 to 2.5 ft). These are exposed along the banks of the Río Tomolosa, 7 km southwest of the city of Tarija. As much as 8 m (26 ft) of river fill commonly buries two or three lignite beds. Farther south, the lignite beds are present across a very broad area. The lignite probably formed in lagoons during interglaciations. It is not useful for domestic consumption, for iron manufacture, and so forth because of its high sulfur content (6 to 8 percent). It could be used as a fuel in the manufacture of bricks. Nearby, similar beds are found close to Padcaya in Quaternary strata. These are more akin to peat and mixed with sand and clay.

In water wells in the Cochabamba basin (Fig. 1), thin lignite beds exist between depths of 3.5 and 7 m (11 to 23 ft). Also near Cochabamba in a small glacial basin 3,500 m in elevation (11,480 ft) near the Patati Ranch in eastern Sabaya, lenticular beds of lignite are present. They are intercalated with sandstone and clay and range in thickness from 5 to 25 cm (0.15 to 0.8 ft). An analysis of these lignites is presented in Table 15. Lignites in the mountains around Cochabamba, examined by Ahlfeld (1954), showed the same quality of beds as in the areas closer to Cochabamba itself.

TABLE 15. CHARACTERISTICS OF PLIOCENE-PLEISTOCENE LIGNITE NEAR COCHABAMBA

Characteristics	Value
Moisture (%)	11.15
Volatile matter (%)	43.46
Ash (%)	30.12
Sulfur (%)	4.20
Btu/lb	4,300

Partly and completely carbonized pieces of tree trunks and branches have been found in the Tertiary Subandine (Tatarenda) Group of the eastern Andes, east of Tarija and elsewhere. Specifically, these clasts of coal are from the origins of Guandacay and the Río Cambarí, Tariquia Department, Arce Province, Tarija; in Ticucha Department, Azero Province, Chuquisaca; and in Villa Tunari (San Antonio), Río Chaparé. The clasts are in red sandstone beds and conglomerates, all of them of fluviatile origin. The original beds from which the clasts were derived have not been found.

This coal is compact and has a bright luster, hygroscopic nature, and conchoidal cleavage. With the microscope, one can observe very clearly the original woody structure. Analyses of the Guandacay coals of the Tarija region are given in Table 16.

Schlagintweit (unpublished report summarized in Ahlfeld, 1954) wrote the following about coal occurrences in the Río Chaparé region of southern and southeastern Bolivia: "Sporadic blocks of carbonized tree trunks are characteristic of the freshwater deposits, of rather turbulent origin, formed in a continental environment. The conservation of the wood and its age must be rather recent. The manner in which the trunks are preserved remind me of the 'jujeño' coal of Salta and Jujuy Provinces [Argentina]. This carbonized wood is found across very great distances.

"I have found in situ coal in the Río Chipiribi, in the Ilibulu Brook, a tributary of the Río Espíritu Santo, and along the Río San Antonio. Some blocks are found isolated along the river beaches. The local inhabitants of course have noticed these and hope to find commercial coal. However, the coal has no economic value."

Resources

From what is now known of Bolivia, there are no commercial resources in the country.

TABLE 16. CHARACTERISTICS OF TERTIARY COALS OF GUANDACAY

Characteristics	Partly carbonized	Completely carbonized
Moisture (%)	13.25	16.23
Volatile matter (%)	23.45	27.77
Fixed carbon (%)	29.10	49.10
Ash (SiO_2) (%)	34.20	6.90
Sulfur (%)	0.95	1.24
Btu/lb	7,050	11,700

ACKNOWLEDGMENTS

The completion of this work was possible thanks to O. Wenzel, Chief Geologist of the Empresa Nacional del Petróleo (ENAP); P. Courard, Manager of Operations, Carbonífera Lota-Schwager S.E., all of whom have given me very valuable information. I also thank the Corporación Minera de Bolivia, which has contributed to me much useful information.

SELECTED REFERENCES

Agassiz, A., 1875-1876, Hydrographic sketch of Lake Titicaca: Philadelphia, Am. Acad. Arts and Sci. Proc., v. II, 200 p.

Ahlfeld, F., 1954, Yacimientos minerales de Bolivia, Chap. 3: Madrid, Imprenta Industrial S.A., p. 252-254.

Angelelli, V., 1941, Los yacimientos minerales y rocas de aplicación de la República Argentina: Buenos Aires, Dirección Minas Geol., Minist. Agricultura de la Nación, Bol. no. 50, 400 p.

Barwick, J. S., 1954, The sub-bituminous coal reserves of Magallanes Province, Chile: Santiago, July 27, Corporación de Fomento de la Producción (unpub. rept.).

Bergmann, F.A.J., 1978, Coal resources of the Argentine Republic, *in* Kottlowski, F.E., and others, eds., Coal resources of the Americas: Selected papers: Geol. Soc. America Special Paper 179 (this volume).

Borrello, A., 1956, Recursos minerales de la República Argentina, Vol. III: Buenos Aires, Inst. Nac. Investigación Ciencias Naturales, Minist. Educación de la Nación, p. 25-316.

Brüggen, J., 1918, Informe sobre el carbón de La Ternera (Copiapó): Santiago, Serv. Geol. Minist. Industria y Obras Públicas, Folleto no. 1, 16 p.

Cortés, R., 1966, Informe sobre las reservas de carbón de Magallanes: Santiago, Empresa Nacional de Petróleo (unpub. rept.).

Forbes, D., 1861, Report on the geology of South America, Pt. I, Bolivia and southern Peru, with notes on the fossils by Huxley, Satter and Jones: Geol. Soc. London Quart. Jour., v. 17, p. 7-62.

Gonzalez, E., 1966, Las reservas de carbón de Magallanes: Santiago, Empresa Nacional de Petróleo (unpub. rept.).

Muñoz C., J., 1946, Estado actual del conocimiento sobre la geología de la Provincia de Arauco: Univ. Chile, Facultad de Ciencias Físicas y Matemáticas, Anales, v. 3, p. 30-63.

Reccius E., J., 1942, Estudio petrográfico de los carbones chilenos, *in* Primer Congreso Panamericano de Ingeniería de Minas y Geología, Anales: Santiago, Imprenta Universo, p. 976-999.

Veijl, C., 1962, Contribución al conocimiento de la geología regional de Concepción: Santiago, Rev. de Minerales, Inst. Ing. de Minas de Chile, Año XVI (January-March), no. 72, p. 21-71.

Wenzel, O., 1955, Las reservas carboníferas nacionales: Rev. de Minerales, Inst. Ing. de Minas de Chile, Año X (January-April), no. 49, p. 49-57.

MANUSCRIPT RECEIVED BY THE SOCIETY DECEMBER 15, 1976
MANUSCRIPT ACCEPTED DECEMBER 21, 1977

Printed in U.S.A.

Geological Society of America
Special Paper 179

Coal resources of the Argentine Republic

Federico A. J. Bergmann
Chacabuco 159, Bernal, Provincia de Buenos Aires, Argentina

ABSTRACT

In general, the coal deposits of Argentina fall into four groups on the basis of their predominant geologic ages: Neogene, early Tertiary, Jurassic, and Triassic. These four groups are related to the tectonic and stratigraphic framework of the country.

Neogene coals are mainly allochthonous lignitic coals of limnic origin; they have large amounts of ash. Predominating in the deposits of early Tertiary age are subbituminous, high-volatile coals of both limnic and paralic origins, locally with some coking properties. The Jurassic and Triassic provinces have bituminous coals with agglomerating and coking properties. The Jurassic coals were deposited mainly in a paralic environment; the Triassic coal-bearing strata mainly represent a limnic environment.

A total reserve of 483,921,000 metric tons is estimated on the basis of very scanty data. Only the Paleogene Río Turbio and Pico Quemado deposits of southern Argentina have been studied in some detail. The coal-bearing early Tertiary Río Turbio deposit contains 99 percent of Argentina's established reserve.

Increased study of the Argentine coals will be pursued as a result of the world energy crisis. It is hoped that much larger reserves than those currently known will be established. During 1974, a total of 625,647 metric tons was produced, and it is hoped that production will be increased to 3,000,000 metric tons per year.

INTRODUCTION

Although interest in the exploitation of Argentine coal had already begun by the beginning of the 19th century, it can be said that more recently—as a result of the severe Argentine energy crisis brought about by World War II—an idea began to crystallize, specifically that of making a sustained and systematic study of Argentina's reserves of combustible fuels. To accomplish this end, the following proposals were made: (1) determine the effects of the energy deficit that was produced by World War II; (2) augment, within reasonable limits, the potential resources that might be developed within the limitations of the technological and economic factors involved in the production of hydrocarbons; (3) make maximum use of petroleum production but maintain a satisfactory balance between petroleum and coal production insofar as fuel-oil requirements are concerned, with the object of producing from each the maximum yields of light-weight and medium-weight products at maximum profit; and (4) build up a raw material base for making metallurgical coke within a reasonable period of time, so that Argentina might be self-sufficient as soon as possible.

In 1941 the Argentine Government through the Yacimientos Petrolíferos Fiscales (YPF) began to carry out the above proposals. The YPF ultimately spawned the Yacimientos Carboníferos Fiscales (YCF) whose specific task was to carry out the exploration for and exploitation, industrialization, and marketing of coals, peat, asphaltite, and oil shale.

Apparently for economic reasons, prospecting for and evaluation of coal reserves began to decline notably toward the end of the 1950s, and as of 1968, coal exploration practically ceased.

In this paper, I try to present a brief sketch of the coal resources of Argentina. This sketch is based on published information and data. I point out that, because of the nature of this type of review, I make reference only to the most important bibliographic references that will be of use to the reader.

COAL PROVINCES

The coal-bearing sequences are exposed particularly in the western part of Argentina in the Andean Cordillera and Precordillera. Despite the considerable size of the areas containing these sequences, it is possible to divide the coals into four groups or provinces (Fig. 1) and to relate them (within the general geologic framework of the country) to the geologic history and fundamental geologic structure of Argentina. The history and structure, in turn, are related directly to the rigid, largely Precambrian cratonic element that forms the central axis of the country—the Central craton (Bracaccini, 1960). This craton may be continuous from north to south through Argentina. East and west around this central platform area are various marginal basins.

Within this broad framework, four coal provinces are identified. These are related directly to specific stratigraphic sequences that contain coal beds, to genetic sequences of deposition in which the coal beds were formed, and to the diagenetic grade of the coals. For each geologic time span and for each area, the coals have certain identifying characteristics. The coal provinces are named after the predominant geologic ages of their coal deposits.

Neogene coals are dominated by lignitic coal beds of limnic

origin. These coals are allochthonous and are characterized by large amounts of diffused ash. In contrast, the coals of the coal-bearing Triassic and coal-bearing Jurassic provinces are bituminous with agglomerating and coking properties, and both are characterized by less intense allochthonous processes. Of the Mesozoic coals, the Triassic coals indicate deposition mainly in a limnic (lacustrine) environment, whereas the Jurassic coals indicate deposition mainly in a paralic (swamp, lagoon, or shallow-neritic) environment. Coals of the coal-bearing early Tertiary (that is, Paleogene, which includes Paleocene, Eocene, and Oligocene) province are high-volatile coals; they are subbituminous or "transitional coals," in the European concept of that term. In the northern part of the area where these coals are exposed, high-volatile bituminous coals are found with coking properties similar to those of the Gulf of Arauco in Chile. This fact may be partly due to a locally high geothermal gradient related to magmatic phenomena. The Paleogene coals are both limnic and paralic and exhibit properties somewhat like those of the Mesozoic coals, particularly with reference to the accumulation and transformation of the vegetal material.

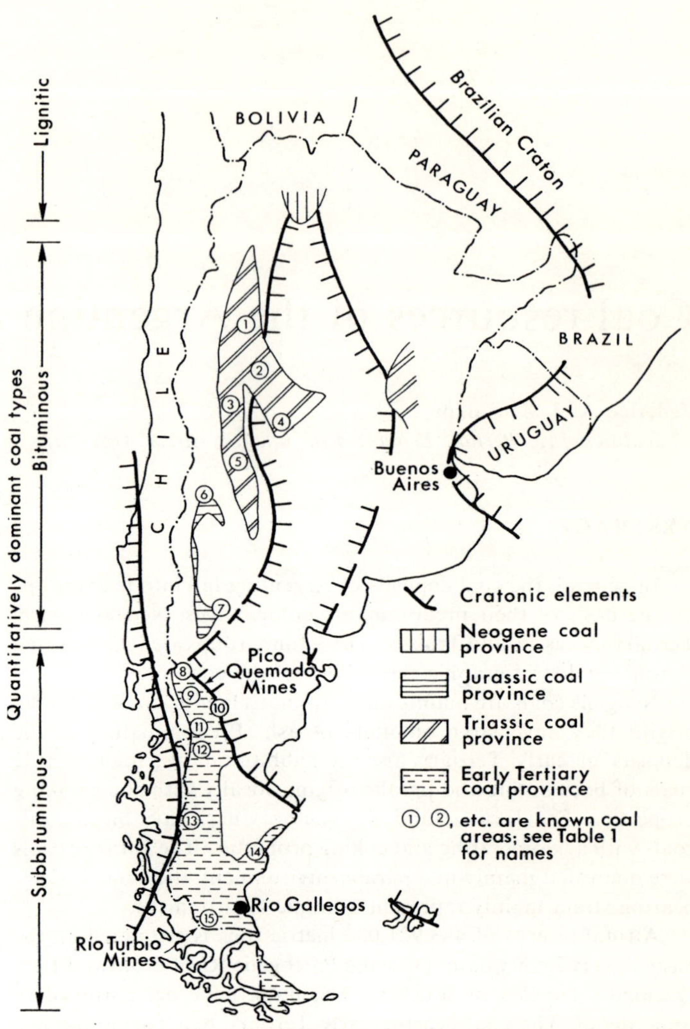

Figure 1. The four coal provinces of Argentina. Most important mining areas are Río Turbio (15) and Pico Quemado (9). Cratonic element west of Buenos Aires and east of the Triassic province is Central craton.

CALCULATIONS OF RESOURCES

Prospecting

Approximately 100,000 km^2 have been covered by geologic reconnaissance, and of this amount, 30,000 km^2 were surveyed with triangulation and topographic mapping. At the same time, about 370 shallow boreholes were drilled, most of them in the Río Turbio coalfield. Stripping and digging operations were conducted concurrently, mainly in the Río Turbio coalfield, for helping in the beginning of exploitation. A valuable reference on this operation, although somewhat outdated, is that by Borrello (1956).

Coal Quantity

Within the four coal-bearing provinces, approximately 250 distinct outcrops and accumulations of coal have been found. Of

TABLE 1. CHARACTERISTICS AND TONNAGE OF COALS STUDIED IN 15 ARGENTINE LOCALITIES

Description of coal	Name of mine	Location no. in Figure 1	Geographical province	Reserves
Semianthracite	Cervantes	6	Mendoza	100,000
Low-volatile bituminous	La Negra	2	La Rioja	55,000
Medium-volatile bituminous	Río Tambillos	1	La Rioja	
Medium-volatile bituminous	Santa Máxima–El Saltito	5	Mendoza	655,000
Medium-volatile bituminous	Burgos	7	Neuquén	
High-volatile bituminous A	La Delfina	3	San Juan	50,000
High-volatile bituminous B	Rickard	4	San Juan	
High-volatile bituminous B	Jorge Newbery	8	Neuquén	
High-volatile bituminous B	Pico Quemado	9	Río Negro	3,100,000
High-volatile bituminous B	Lepá	12	Chubut	
High-volatile bituminous C	La Criolla	13	Santa Cruz	70,000
Subbituminous A (partly high-volatile bituminous C)	Río Turbio	15	Santa Cruz	478,976,000
Subbituminous A/C	Indio	10	Chubut	
Subbituminous A/C	Santa Ana	11	Chubut	915,000
Subbituminous A/C	Cabo Curioso	14	Santa Cruz	
			Total	483,921,000

the 250, a total of 15 deposits has been partly investigated in some degree of detail. The investigations of these 15 were made according to the standards of the United States Geological Survey and the Bureau of Mines of the United States.

Table 1 summarized the total amount of coal found in the 15 deposits that have been partly excavated. Classification of the coals is made according to the standards set by the American Society for Testing and Materials.

Only the Río Tambillos, Pico Quemado, Río Turbio, Indio, and Cabo Curioso mines include resources in the proved and probable as well as the potential categories. As can be seen from Table 1, practically all of the known coal is in the Río Turbio coalfield. More than half of the remaining (0.55 percent of the total) comes from the Pico Quemado mine where a resource of 2,640,000 metric tons (hereafter, t) of coal has been determined. The remainder (0.45 percent) is distributed among the other 13 deposits.

CHARACTERISTICS OF THE COALS

Composition

Table 2 shows the estimated percentage of different grades of coal in Argentina and the estimated ash and moisture contents of these coals. Table 3 shows the principal characteristics of cleaned coals from Río Turbio and Pico Quemado mines. I provide data only for these two mine areas because together they contain almost all of the known reserves of Argentina.

Geologic Distribution

Figure 2 shows the variations of some basic characteristics of the coals from each of the four coal-bearing provinces. Among the parameters portrayed on Figure 2, water was selected because of its connection with the structural aspects of coaly materials. Figure 2 shows a direct relationship between the geologic characteristics of each coal province and the metamorphic grade of the coals. I call particular attention to the Tertiary coals which have lower moisture values and which in places have some agglomerating properties (see Tables 2 and 3).

Economic Evaluation of Río Turbio Deposits

In the mid-1960s, the Argentine Government contracted the services of a mining and economic minerals consulting firm, Powell Duffryn Technical Services, Ltd., of London, and asked them to study the economic feasibility of exploiting the Río Turbio mines where, as I have noted, the greatest coal reserves of Argentina are concentrated. These coals belong to the Paleogene group (early Tertiary).

The study of this consulting firm reached the conclusions that with a commercial production of about 1,200,000 t/yr, a good economic and financial balance would be reached and that the coal produced would be competitive in the internal Argentine market which is concentrated around Buenos Aires. (The conclusions of the firm can be seen in the following public documents: Memoria Anual y Balance General de YCF of 1967; Press Conference of the Administrador de YCF, September 4, 1968.) The conclusions of the consulting firm should be qualified with the observation that, at the

TABLE 2. PERCENTAGE OF COAL TYPES, ASH CONTENT, AND MOISTURE CONTENT OF ARGENTINE COALS

Type of coal	Reserve (%)	Ash content (%)	Moisture content (%)
Semianthracite	0.02	26	2
Low-volatile bituminous	0.01	28	2
Medium-volatile bituminous	0.13	31	4
High-volatile bituminous A/C	0.67	25	4
Subbituminous A (in part high-volatile bituminous C)	98.97	30	12
Subbituminous A/C	0.20	34	12

TABLE 3. ANALYSES OF RIO TURBIO AND PICO QUEMADE COMMERCIAL COALS

Characteristic measured	Río Turbio	Pico Quemado
Proximate analysis		
Moisture	11%–14%	3%–5%
Ash (dry)	10%–25%	10%–20%
Volatile matter (dry, ash-free)	46%–50%	42%–46%
Fixed carbon (dry, ash-free)	50%–54%	54%–58%
High calorific value (dry)	5,800–7,000 kcal/kg	6,300–7,100 kcal/kg
Element (ultimate) analysis (dry, ash-free)		
Carbon	75%–79%	73%–82%
Hydrogen	5%–8%	5%–7%
Nitrogen	0.5%–0.15%	0.7%–0.8%
Total sulfur	0.5%–1.1%	0.2%–0.5%
Oxygen	11%–16%	10%–19%
Free swelling index*	0-0.5%	5%
Fusibility of ash		
Softening point	1250°–1280° C	1400°–1568° C
Fusion point	1305°–1405° C	1570°–1600° C
Liquefying point	1340°–1500° C	1600°–1628° C
Petrographic composition†		
Vitrain (vitrinite)	20%	60%
Clarain (clarinite)	70%	25%
Durain (durinite)	5%	11%
Fusain (fusinite)	5%	4%

*In dry coal with 10% ash.
†Macerals, cleaned coal; values are percentage volumes.

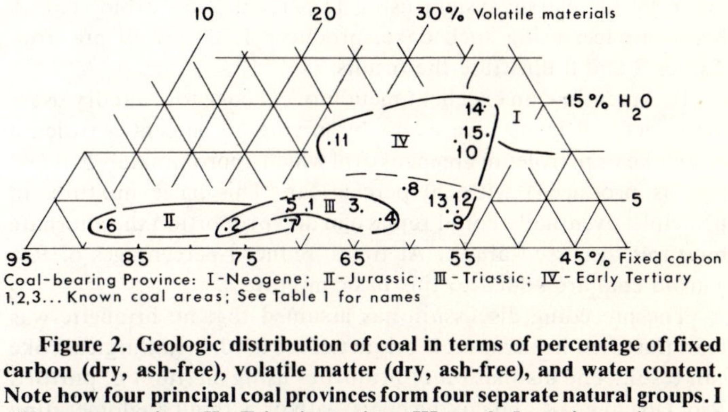

Figure 2. Geologic distribution of coal in terms of percentage of fixed carbon (dry, ash-free), volatile matter (dry, ash-free), and water content. Note how four principal coal provinces form four separate natural groups. I = Neogene province; II = Triassic province; III = early Jurassic province; and IV = Paleogene (early Tertiary) province. Arabic numbers refer to coal outcrops and mine enumerated in Table 1.

time the study was made, only about 20 percent of the Río Turbio coals, reachable by drift and slope mining, was workable from an economic viewpoint.

POSSIBILITIES OF CLEANING AND THEIR APPLICATIONS

In the YCF laboratories, as well as in specialized laboratories of Germany, France, and the United States, research investigations and experiments were carried out using Río Turbio coal. The following is a summary of the principal results (Bergmann, 1963), to which I have added some remarks about the other coals.

Purification

The high ash content (due in part to mechanical processes), the fact that the ash content is dead weight in transportation, and marketing standards require that the Río Turbio coal be cleaned. The classic system of gravity cleaning has been adopted to improve the qualities of this coal. The washability curves indicate the presence in the coal, in a diffused state, of significant amounts of ash. Table 4, designed for general orientation purposes, shows some outputs of the commercial sizes.

Coking

Table 5 shows the average distillation outputs on pure grades of coal. The solid residues are pulverized, and the investigations and experiments carried out were done with the objective of maximizing the utilization of this coal in the manufacture of conventional metallurgical coke. Such investigations considered distillation in one and two stages.

Single-Stage Coking. This technique, widely developed in the industry, includes variations of classic treatments recommended for noncoking coals. The experiments were developed on a pilot scale in France and were pursued further in Argentina (Barreras, 1965; Filippini, 1966). In general terms, good metallurgical coke is obtained by adding to a typical coking mixture—consisting of 70 percent high-volatile coal and 30 percent low-volatile coal—increasing amounts of Río Turbio coal as shown in Table 6.

In the blast furnace of the Sociedad Mixta Siderurgia Argentina (SOMISA) in San Nicolás, Buenos Aires Province, coke is made by the classic system using 15 percent Río Turbio coal. A blast furnace, using such coke, produces 1,500 t/d of pig iron. Tables 7 and 8 illustrate the results.

In Argentina, an excellent metallurgical coke for foundry use is manufactured in beehive ovens by mixing 80 percent petroleum semicoke or petroleum greencoke (of which approximately 500,000 t/yr is produced) with 20 percent tar. This basic mixture, in principle, eventually could replace in any proportion the alternate mixtures discussed above. At times, reduced percentages of Río Turbio coal are added to this basic mixture.

The preceding discussion has assumed that no briquette was manufactured in advance. If briquettes are used, metallurgical coke can result. The distillation of briquettes using mixtures of purified Río Turbio coals and coking coals, with or without agglomerating agents, yields metallurgical cokes. The amount of Río Turbio coal which can be added ranges from 25 to 50 percent. The amount that is added depends on the mixture of other ingredients and on the briquette pressure, which was proved up to 2,500 kg/cm^2.

Double-Stage Coking. To this point I have discussed only a single-stage method to manufacture coke. A double-stage method also can be used. In this technique, the coal first must be degasified. This is necessary if acceptable metallurgical cokes are to be manufactured. These appear in the form of briquettes consisting of

TABLE 4. RELATIONS AMONG DENSITY, SIZES, AND ASH CONTENT OF COALS, RIO TURBIO

Density	Flotation accumulations (<0.5 in.)	Flotation accumulations (0.5–3 in.)
1.35	40%, with 8% ash (dry)	40%, with 11% ash (dry)
1.60	70%, with 15% ash (dry)	63%, with 17% ash (dry)

TABLE 5. DISTILLATION OUTPUTS ON CLEANED COAL, RIO TURBIO

Distillation temperature (° C)	Semicoke/coke (%)	Anhydrous tar (%)	Liquids (%)	Gas (%)
550	64	19	8	9
1000	60	7	14	19

TABLE 6. RELATIONS AMONG GRAIN SIZE, MOISTURE CONTENT, AND CHARGE DENSITY, RIO TURBIO COAL

Río Turbio coal (%)	Grain size (mm)	Moisture content (%)	Bulk density (kg/m^3)
15	6	8	700*
20	2†	8	780
25	2	4§	850
30–35	2	4	1,000#

*Classic system.
†Methodic grinding.
§Natural moisture.
#By tamping.

TABLE 7. DIFFERENT COALS UTILIZED IN MANUFACTURE OF COKE, SAN NICOLAS, ARGENTINA (SOMISA)

	Types of coal		
	High-volatile (USA) (%)	Low-volatile (USA) (%)	Río Turbio (%)
Moisture content	7.3	5.9	10.6
Volatile matter	29.3	16.0	39.0
Ash content	3.5	5.1	9.5
Total sulfur	0.92	0.96	1.03
Free swelling index	9	8.5–9	0.5

TABLE 8. TYPES OF COKE RESULTING FROM DIFFERENT COMBINATIONS OF COALS SHOWN IN TABLE 7

Properties of coke	Coal mixture 1	Coal mixture 2
Granulometry, less than 1 in.	1.4%	1.7%
Shatter index, more than 2 in.	84.9%	88.0%
Stability index	65.1%	59.7%
Hardness index	74.1%	65.3%

Note: Coal mixture 1 is 70% high-volatile (USA)–30% low-volatile (USA) coals. Coal mixture 2 is 60% high-volatile (USA)–25% low-volatile (USA)–15% Río Turbio coals.

semicoke or coke cemented together with derivatives of primary and secondary tars. In one method of making these briquettes, it is necessary to add 20 to 30 percent of coking coal. In a second method, the process is completed satisfactorily by using only coal from Río Turbio. However, both methods are costly and thus useful only in special situations (such as a protected or closed economy). Currently, the first method is being readjusted for use in continuing processes so it will have competitive commercial value. In the case of Río Turbio coal, the double-stage technique encounters a problem because the ash content naturally increases as the coke is formed.

Gasification and Synthesis

Between 1949 and 1954, the possibilities of gasification and synthesis were studied according to different methods in Germany and in the United States.

Lurgi-Ruhrchemie (Germany) Process. The first method developed consumes only coal of 6,000-kcal/kg quality at a rate of 3,000 t/d. Production of synthesis gas requires coal consumption of 2,400 t/d. The following are derivatives from the tars obtained during the production of synthesis gas: tars and oils, 204 t/d; raw benzine, 51.5 t/d; and phenic (carbolic) oil, 13.7 t/d. The products of synthesis-gas production by the Fisher-Tropsch method at a pressure of 20 kg/cm^2 are the following: solid wax, 93.7 t/d; soft wax, 75 t/d; heavy oil, 75 t/d; medium oil, 75 t/d; light oil, 38.4 t/d; liquid gas, 11.3 t/d; and alcohols, 7.9 t/d.

In an alternative method, coal of 5,500-kcal/kg quality is used at a rate of 2,000 t/d. For producing synthesis gas, coal consumption is 1,400 t/d. The products obtained during gasification, plus the products formed during synthesis by means of the Fisher-Tropsch method at medium pressure are the following: gasolines, 224.4 t/d; gas-oil, 28.2 t/d; diesel oil, 61.3 t/d; fuel oil, 11 t/d; and tar-pitch, 56.6 t/d.

Hydrocarbon Research, Inc. (USA), Process. One method utilizes coal of 6,000-kcal/kg quality at a rate of 5,100 t/d. To produce synthesis gas, 4,400 t/d of coal are needed. The products of Hydrocol synthesis are gas of 3,420-kcal/m^3 quality, 716,400 m^3/d; gas of 18,500-kcal/m^3 quality, 19,600 m^3/d; catalytic-polymerization oil, 314.8 m^3/d; treated synthetic oil, 686 m^3/d; tars from polymerization, 22.1 m^3/d; tars and light oils of distillation, 467 m^3/d; gas-oil, 152.3 m^3/d; residual paraffin, 16.7 m^3/d; and oxygenated chemical products, 140 kg/d.

In an alternative method, coal of 6,000-kcal/kg quality is consumed at the rate of 5,400 t/d. Coal consumption in order to produce synthesis gas is 4,680 t/d. The products of Hydrocol synthesis are gas of 3,400-kcal/m^3 quality, 716,400 m^3/d; gas of 22,800-kcal/m^3 quality, 91,000 m^3/d; catalytic-polymerization oil, 314.8 m^3/d; treated synthetic oil, 686.5 m^3/d; motor oil, 779 m^3/d; and oxygenated chemical products, 140 kg/d.

Hydrogenation

Because of the high ash content of the coal, experimental hydrogenation was carried out on the tars and light oils produced in the above-described processes. Hydrocarbon Research, Inc., treated these products (tars and light oils) with hydrogen (96 percent) obtained from the synthesis gas. These experiments produced similar quantities and qualities of products. For example, the use of the equivalent of 580 m^3/d of tars and oils produced the equivalent of 71,000 m^3/d of gas of 24,000-kcal/m^3 quality and 676 m^3/d of oil of 62 to 64 octane.

According to the results of new processes (for example, using fluidized coal beds with catalyzing agents), it would be very interesting to see what possibilities would result from the application of such processes to the Río Turbio coals. These coals are in a coal-bearing region that also contains natural gas. Such conditions suggest the possibility of obtaining hydrogen (H_2) through direct processes or through industrial-recovery methods.

Combustion

On the basis of the characteristics of the coal—which gives off a theoretical combustion temperature between 1600° and 1250° C for 10 and 25 percent ash (dry) contents, respectively—it can be said that there are no technical limitations for using this coal for industrial purposes. For industrial installations that are designed for coal containing less volatile material and less ash, one may resort to the mixtures of Río Turbio coal and petroleum semicoke or petroleum greencoke that were mentioned above.

The coal of coarse size (greater than 0.5 in.; see Table 4) has little grain-size stability. This difficulty in some cases can be overcome by grinding it to a finer size and even making briquettes with 7 percent asphalt from Argentine-produced petroleum used as an agglomerating agent.

Pico Quemado Coal and Other Coals

The attributes of the Pico Quemado coal are similar to those of the Río Turbio coal, except for the properties of fusion and agglomeration. Because of its promising coking properties, the Pico Quemado coal was tested on a pilot scale at the coke ovens of SOMISA. These studies showed that with 10 percent ash (dry basis), the Pico Quemado coal can replace from 30 percent and possibly as much as 60 percent of the basic mixture of coking coals that are imported.

The remaining Argentine coals are little known with reference to quantity and quality. They can be considered as having potential for their bituminous properties as well as for their coking properties, but they still remain to be investigated. The few investigations made to date indicate that these coals have a rather high ash content, but it may be presumed in general that purification of the coals could give a better yield than would purification of the coal from Río Turbio, particularly in the case of the bituminous coals because of their higher ash content.

PRESENT MINING OPERATIONS

The only major operations presently in progress are those at Río Turbio and, to a certain extent, Pico Quemado. The remaining coals should be considered to be only dispersed outcrops which are not economic at the present.

Río Turbio Deposit

The Río Turbio coalfield is centered at lat 51° 35′S, long 72° 17′W. It is 260 km from the Atlantic coastal port of Río Gallegos and 2,000 km from Buenos Aires.

The elevation of the Río Turbio deposit is 250 m above sea level. The coal-bearing strata belong to the Cerro Dorotea Group of Paleogene age which overlies an Upper Cretaceous marine succession. This is the Cerro Cazador Group which contains ammonites (Fig. 3).

The Paleogene Cerro Dorotea Group contains some coal beds scattered through a thickness of 1,000 m. The entire succession is an alternation of marine and continental strata. It contains *Nothofagus* sp. associated directly with the coal beds. The overall environment of deposition of the Cerro Dorotea Group can be characterized as paralic.

In general, the lower part of the Paleogene succession has been poorly explored. Within it, some coal beds contain as much as 25 percent volatile matter (pure coal basis) with some coking properties. Near the middle part of the Cerro Dorotea Group (and separated stratigraphically by 280 m) are two zones bearing most of the coal.

The upper zone contains 377,000,000 t of raw coal in three beds. The Dorotea bed is uppermost and bears 72 percent of the total volume of coal; the bed "A" in the middle contains 23 percent, and the "B" bed at the bottom of the upper zone contains 5 percent. The thickness of the Dorotea bed ranges from 1.2 to 2.0 m. The thickness of bed "A" ranges from 0.7 to 2.2 m. The thickness of bed "B" ranges from 0.5 to 1.0 m. The stratigraphic separation between the Dorotea and the "A" bed is about 8 m, and that between the "A" and the "B" beds is about 15 m.

The lower zone contains 101,000,000 t of raw coal. This zone is divided into two beds: the Superior bed with 47 percent of the tonnage and the Inferior bed with 53 percent of the tonnage. The Superior bed ranges in thickness from 1.8 to 2.6 m, and the thickness of the Inferior ranges from 7.0 to 1.0 m. The exploitable thickness actually does not exceed 1.5 m. The stratigraphic separation between these two beds is approximately 12 m. The continuity of this zone north of Río Turbio mine no. 3 still has to be defined with drilling and stripping.

In the upper part of the productive section, coal marker beds have been found. The upper part of the Cerro Dorotea Group grades into Neogene sedimentary rocks of continental facies that are divided into the Río Turbio and Latorre Groups.

All of the strata in the area are gently inclined with dips on the order of 10°. For all intents and purposes, they are almost unfractured.

Mining activity, carried out by Yacimientos Carboniferos Fiscales (YCF), is concentrated mainly in the Dorotea bed, especially in mines nos. 3 and 4 (Fig. 3). Mines nos. 3 and 4 have been developed by level and inclined drifts that have made it possible to mine the coal by the "overhanging method," that is, with large faces that can be excavated as much as 160 m back from the galleries. Special mining equipment is employed that permits the use of continuous miners — both of the boring type and of the drum type. Along the mined faces, both hydraulic safety posts and roof supports are used; the supports may be either of timber or of metal, according to the conditions encountered. The transport of the coal is by means of conveyor belts and electrically driven carts. Production as of 1968 was on the order of 450,000 t/yr (see Addendum for more recent information). The yield of raw coal per mining man-face and man-total is about 9 and 2.5 t/d, respectively.

The coal currently being mined contains 35 to 40 percent ash. This coal is treated for the removal of ash in a plant constructed by the Skoda firm (Czechoslovakia). The plant has a treatment capacity since 1973 of 500 t/h. Before 1973, the capacity was only 250 t/h. The coal thus treated is 40 percent coarse size (0.5 to 3 in.) and 60 percent fine size (less than 0.5 in.). The coarse coal is put in containers of medium density in a watery suspension containing magnetite. The smaller sized coal is put in feldspar jigs. The rate is 60 t/h. At present, the wash-efficiency is 60 percent.

The mixed grades of coal are burned by YCF to produce electrical energy. There are three power plants: one of 6,000 kW and two of 7,500 kW each. The electricity produced by this means supplies the mines and the local population, which consists of about 5,000 people whose livelihood depends directly or indirectly on mining activity.

Most of the commercial coal is transported to Río Gallegos on a 75-cm-gauge railroad. Some of this coal is then unloaded in a delivery yard which holds 30,000 t. It is transported to local consumers by trucks. Most of the coal, however, is loaded directly onto ships by means of a mechanical system with a 1,000-t/h capacity (600 t/h until 1969).

The ships are both government and commercially owned. Because the tides are about 10 m and the construction facilities are inadequate in the port of Río Gallegos, the ships that carry the coal are of small capacity — 3,000 to 5,000 t. It is hoped that with some port improvements, ships with capacities of 8,000 t will be able to operate there. One facility of this type was being constructed in 1968.

Approximately 80 percent of the coal produced is sold in Buenos Aires and vicinity. It is used especially in electric power houses (SEGBA, AyEE, F. C. MITRE), in the coking plant of SOMISA, and in minor industries.

YCF — the operator of the Río Turbio mines — originally planned to triple production by 1972 to a minimum of 1,200,000 t/yr, but this production was not attained (see Addendum). The market for such production is assured, particularly in thermoelectric plants and coking facilities. In the coking facilities, the possibility exists of using the Río Turbio coal mixed with petroleum greencoke that is also produced in Argentina.

Pico Quemado Deposit

The Pico Quemado coalfield is located at lat 41° 35′S, long 70° 53′W, in Norquinco Department, Río Negro Province, 70 km south of the railroad station of Ñirihuao, near the town of Bariloche. The Pico Quemado deposit is at an elevation of 1,300 m. The deposit is in a tectonic erosional remnant — specifically, a syncline within the rigid platform underlain by the Andesite series ("Serie Andesítica"; Fig. 4). The coal is in two productive zones separated stratigraphically by 35 to 70 m. Each productive zone consists of two coal beds, 1 to 2 m thick. Recent studies (Bergmann, 1966) have shown that the sedimentary rocks at Pico Quemado belong to the lower part of an 800-m-thick unit called the Pico Quemado complex. The type locality of the complex crops out a short distance north of the coal deposit. There are some outcrops of coal in the type locality.

Regional geologic studies also show that unconformably above the Pico Quemado complex are the beds of the Arroyo Chenqueniyeu complex, associated with the Patagoniense series (Oligocene). The Arroyo Chenqueniyeu complex also contains coal which is

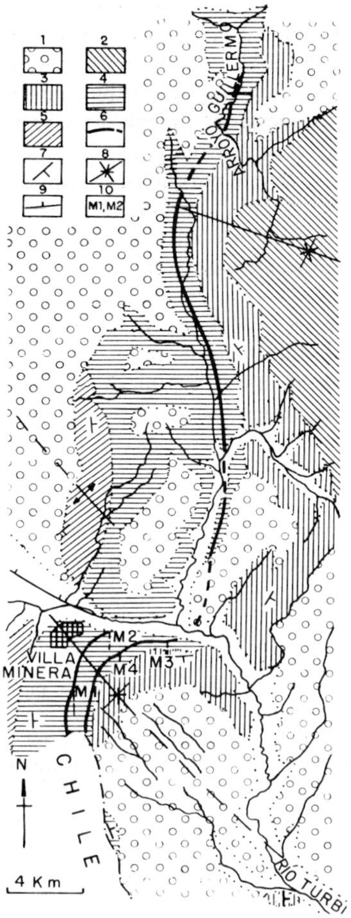

Figure 3. Geologic sketch map of Río Turbio deposits: 1, glacial Pleistocene deposits; 2, Latorre Group (Neogene continental deposits, with some pyroclastic materials); 3, Río Turbio Group (Neogene sedimentary rocks); 4, Sierra Dorotea Group (Paleogene marine and continental sedimentary rocks with coal); 5, Sierra Cazador Group (Upper Cretaceous rocks in marine facies); 6, line of outcrop of stratigraphically highest group of coals (Dorotea, "A," and "B" coal beds) and lower zone (Superior and Inferior beds); 7, strike and dip symbol; 8, syncline symbol (oppositely pointing arrows indicate an anticline); 9, faults; 10, local mine numbers.

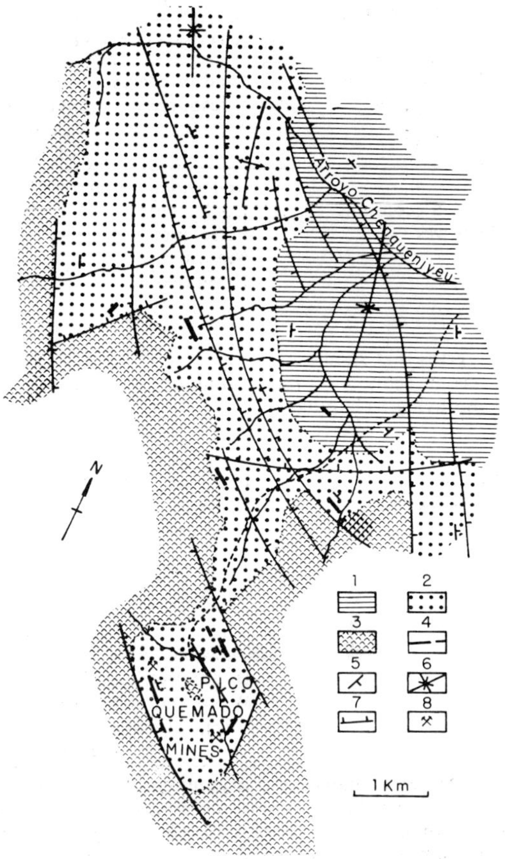

Figure 4. Geologic sketch map of Pico Quemado coal deposit and environs. All rocks appear to be of Paleogene age. 1, Arroyo Chenqueniyeu complex; 2, Pico Quemado complex; 3, Andesite series ("Série Andesítica"); 4, coal outcrop; 5, strike and dip symbol; 6, syncline symbol (oppositely pointing arrows indicate an anticline); 7, faults; 8, mine.

partly of Oligocene age. The coal beds here may be related in age to the Jorge Newbery, Lepá, and other coal deposits of the northern early Tertiary coal province (Fig. 1; Table 1).

The Pico Quemado coal deposit is owned by the Compañía Minera e Industrial de Río Negro S.R. Ltda. (Sociedad de Responsabilidad Limitad), which at the moment is temporarily mining small amounts of coal, without purifying it, and is selling it locally.

If the Pico Quemado deposit has larger potential than it now appears to have, the fact should be pointed out that, for the purposes of the internal Argentine market, this coal possesses certain valuable coking properties. However, this favorable aspect of the coal must be balanced by the fact that the reserves are not yet known to be sufficient to encourage a thorough mining development on a commercial scale.

ESTIMATES OF ROLE OF COAL IN TOTAL ENERGY POTENTIAL OF ARGENTINA

The potential role of coal and other energy resources in Argentina is summarized in Table 9. From the data given above, and especially from the data given in Table 9, it is apparent that there are theoretical limits beyond which resources will not last. The length of time can be deduced by the following equation:

$$A + Bx = C \int_0^x Dx \, dx,$$

where A is the nonrenewable resource potential, B is the renewable resource potential, C is the total energy consumption in Argentina during the base year, 1967 (24,835,000 t of equivalent petroleum), and D is the total rate of increase of annual energy use for the period of x years.

If one takes into account only the known economic reserves of nonrenewable resources and the renewable resources (Table 9) and if one also assumes an increase in annual energy use of $D = 4$ percent (the same increase that has characterized the history of Argentina), then the value of x is 21 yr. If one assumes that $D = 7$ percent, then $x = 17$ yr. These two durations indicate in principle and within a definite frame of reference the true value of the national reserves. The facts given show that the future of Argentina's reserves depend on market requirements, as well as the development of changing technology. Thus, there is considerable uncertainty in these estimates. The uncertainty pertains particularly to coal — not only in Argentina, but everywhere in the world (Mainguy, 1967). The future utilization of coal will depend on advances in nuclear energy and the future structuring of the energy market with reference to the raw materials that are used to produce energy.

CONCLUSIONS

It can be seen that Argentina's future coal potential, from the known facts, is rather modest. Because of this, Argentina has been obliged to develop this resource in a way that will not compromise

TABLE 9. EVALUATION OF ENERGY RESOURCES OF ARGENTINA (1968)

Type of reserve	Kind of energy resource	Factors involved in calculation of value*	Value†
	Nonrenewable resources		
Known economic reserves	Coal	1, 2, 3	20
	Petroleum	1, 4, 5	280
	Natural gas	1, 3, 5	150
	Uranium§	6, 7, 8	35
		Subtotal	485
Reserves that should become economic with improving technology	Coal	1, 2, 3	82
	Uranium#	8, 9	450
		Subtotal	532
Potential future reserves	Peat	1, 3, 5	34
	Oil shale	1, 5, 10	52
	Uranium**	8, 9	600
		Subtotal	686
		Total nonrenewable resources	1,703
	Renewable resources		
	Hydroelectric	7, 8, 11	14
	Wood and other vegetal matter	1, 3	2
		Total renewable resources	16

*1, Mineral volumes or consumptions (for example, of wood) according to 1968 official data; 2, mining and coal washer output; 3, combustible calories with reference to the base 10,500 kcal/kg; 4, specific gravity of 0.9 for petroleum; 5, loss occurs in production, transport, or treatment; 6, installed capacity for electricity; 7, utilization factor of 0.75; 8, 2,200 kcal/kWh/10,500 kcal/kg; 9, as much as 7,000 MW/d/t U; 10, 7% of oil (in oil shale); and 11, useful power.

†Value is given in 10^6 t of equivalent petroleum at 10,500 kcal/kg.

§Assumed to be equivalent to 1,000 MW of electricity, maintained for 25 yr with a concentrated (75%) base of U_3O_8 (at a cost of U.S. $5 to $10/lb).

#Assumed to be equivalent to 13,000 MW of electricity, maintained for 25 yr with a concentrated (75%) base of U_3O_8 (at a cost of U.S. $15/lb).

**Assumed to be equivalent to 17,000 MW of electricity, maintained for 25 yr with a concentrated (75%) base of U_3O_8 (at a cost of U.S. $30/lb).

future flexibility until some major change in energy base occurs. New developments may take place as they have at Río Turbio; these could expand the energy base of Argentina.

ADDENDUM

Since the preceding pages were written in 1968 (and subsequently updated only in part), few things have changed in Argentina. However, the energy crisis of the 1970s has modified the market conditions in Argentina and has brought about the adoption of new policies directed toward petroleum substitutes. For example, one of the Argentine goals is to develop to the maximum amount its hydroelectric potential; such development has nearly tripled during the last few years. Also, the Argentine Government is trying to develop nuclear energy and energy derived from coal.

With reference to coal, the reserves summarized in the preceding pages remain the same, because coal exploration only recently has begun to revive after many years of inactivity. The principal mining of coal continues to be concentrated in the Río Turbio district. Production there during 1974 was 625,647 t. Plans now exist to increase this production to an annual rate of 3 million t as soon as possible. Almost all of the additional coal to be mined will be used for the generation of electricity along the interior streams and along the tidal areas of the Argentine coast.

REFERENCES CITED

Barreras, Armando D., 1965, La elaboración de coque metalúrgico en la Sociedad Mixta Siderurgia Argentina (SOMISA): Bogotá, Sexto Cong. Latinoam. Siderurgia (6th Latin American Siderurgical Cong.), Inst. Latinoam. del Hierro y del Acero, Memoria Técnica, Carbón; Minería del Fierro, v. 3, p. 112-145.

Bergmann, Federico A. J., 1963, Recursos de carbón y otros combustibles sólidos minerales de la República Argentina: Buenos Aires, Yacimientos Carboníferos Fiscales Press, 37 p. (proc. vol., 10th Sess., Comisión Económica para América Latina).

——1966, Contribución al conocimiento de los depósitos con carbón de la zona de Pico Quemado (Provincia de Río Negro): La Plata, Argentina, Univ. Nac La Plata, Facultad de Cienc. Nat. y Mus., unpub. rept., 42 p.

Borrello, Angel V., 1956, Recursos minerales de la República Argentina. III, Combustibles sólidos minerales: Buenos Aires, Rev. Inst. Nac. Inv. Cienc. Nat., 665 p.

Bracaccini, Osvaldo, 1960, Lineamientos principales de la evolución estructural de la Argentina: Buenos Aires, Inst. Argentino Petróleo, Petrotecnia, v. 10, no. 6, p. 57–69.

Filippini, José Rosario, 1966, El carbón argentino, sus posibilidades en siderurgia: Bogotá, Sexto Cong. Latinoam. de la Siderurgia (6th Latin American Siderurgical Cong.), Inst. Latinoam. del Hierro y del Acero, Memoria Técnica, Carbón; Minería del Fierro, v. 3, p. 85–110.

Mainguy, Y., 1967, L'économie de l'énergie: Paris, Ed. Dunod, Colln. Finance et Economie Appliquée, v. 31, 532 p.

MANUSCRIPT RECEIVED BY THE SOCIETY OCTOBER 9, 1975
MANUSCRIPT ACCEPTED OCTOBER 30, 1975

Printed in U.S.A.

Geological Society of America
Special Paper 179

Coal in Brazil

Eurico Rômulo Machado
Companhia Riograndense de Mineração, Rua Botafogo 610, Porto Alegre, Rio Grande do Sul 90,000, Brazil and Department of Geology, Universidade Federal de Rio Grande do Sul, Brazil

ABSTRACT

Brazil has five coal-bearing regions: (1) Pliocene(?) coal in the Pébas Formation of the Upper Amazon basin; (2) Eopaleozoic coal in the Rio Fresco Formation of the Rio Fresco region; (3) Pennsylvanian coal in the Piauí Formation of the Tocantins-Araguaia region, western Parnaíba basin; (4) Mississippian coal in the Potí Formation, eastern Parnaíba basin; and (5) Late Carboniferous or Early Permian coal in the Paraná basin. The Paraná basin contains all of the country's commercial coal, which is in the Rio Bonito Formation of the Gondwana sequence. Total blocked reserves are 3.2 billion metric tons, of which about 5.6 million tons of run-of-mine coal is extracted annually.

INTRODUCTION

The purpose of this paper is to present an overview of known Brazilian coal resources. With the exception of some Tertiary lignite deposits of limited significance, the most important coal occurrences are in the following areas (Fig. 1).

I. Upper Amazon Region

Within the vast Upper Amazon sedimentary basin, lignite seams are interbedded in the upper part of the Pébas Formation, believed to be of Pliocene age. Because of the large area underlain by the Pébas Formation, the Comissão do Plano de Carvão Nacional (CPCAN = National Coal Planning Commission) decided in 1966 to map the surface geology of the area and to drill five test holes. The drilling revealed the presence of several lignite beds in the upper part of the Pébas Formation. However, the lignite is of inferior quality, and the beds are less than 1.5 m thick. Additional work was done in the area as a part of the RADAM Project, and modern airborne geophysical techniques were utilized. The RADAM study was able to define more precisely the distribution of the Pébas Formation and made it possible for Federal agencies (Department Nacional da Producão Mineral = DNPM; Companhia de Pesquisa de Recursos Minerais = CPRM), as well as mining companies, to continue prospecting in this area. All prospecting to date has been of a preliminary nature.

II. Rio Fresco Region

Layers of anthracite-like coal are interbedded with the Rio Fresco Formation. The age of the formation is unknown, but appears to be Eopaleozoic. The coal beds are less than 1 m thick. The Rio Fresco Formation is in a sedimentary basin that was formed on the top of the crystalline Brazilian shield in the central part of Brazil, particularly in the state of Pará. The coal occurrences were studied by the Instituto de Desenvolvimento do Estado do Pará (IDESP) under the auspices of CPCAN. To date, no coal has been found that can be developed economically.

III. Tocantins-Araguaia Region

The region between the Tocantins and Araguaia Rivers lies on the western flank of the Parnaíba sedimentary basin and occupies parts of the states of Goiás, Maranhão and Pará. Thin lenses of excellent-quality coal are present within the Pennsylvanian Piauí Formation. The coals are associated with a Westphalian-Stephanian flora. The coal occurrences led CPCAN in 1968 to begin geologic studies and drill nine test wells. The results were disappointing, because no economic deposits were found.

IV. Western Piauí Region

Coal has been found along the eastern flank of the Parnaíba basin. Thin beds (maximum thickness about 10 cm) of high-quality coal, associated with a boreal Mississippian flora, have been found in the upper part of the Potí Formation. The lower part of the Potí is marine and is of Early Devonian age. The coal is interbedded with shale and siltstone and associated with lesser amounts of sandstone. Field studies and drilling were carried out by the DNPM and by Petróleo Brasileiro S.A. (Petrobrás). To date, none of the studies — including 20 test holes — has revealed the presence

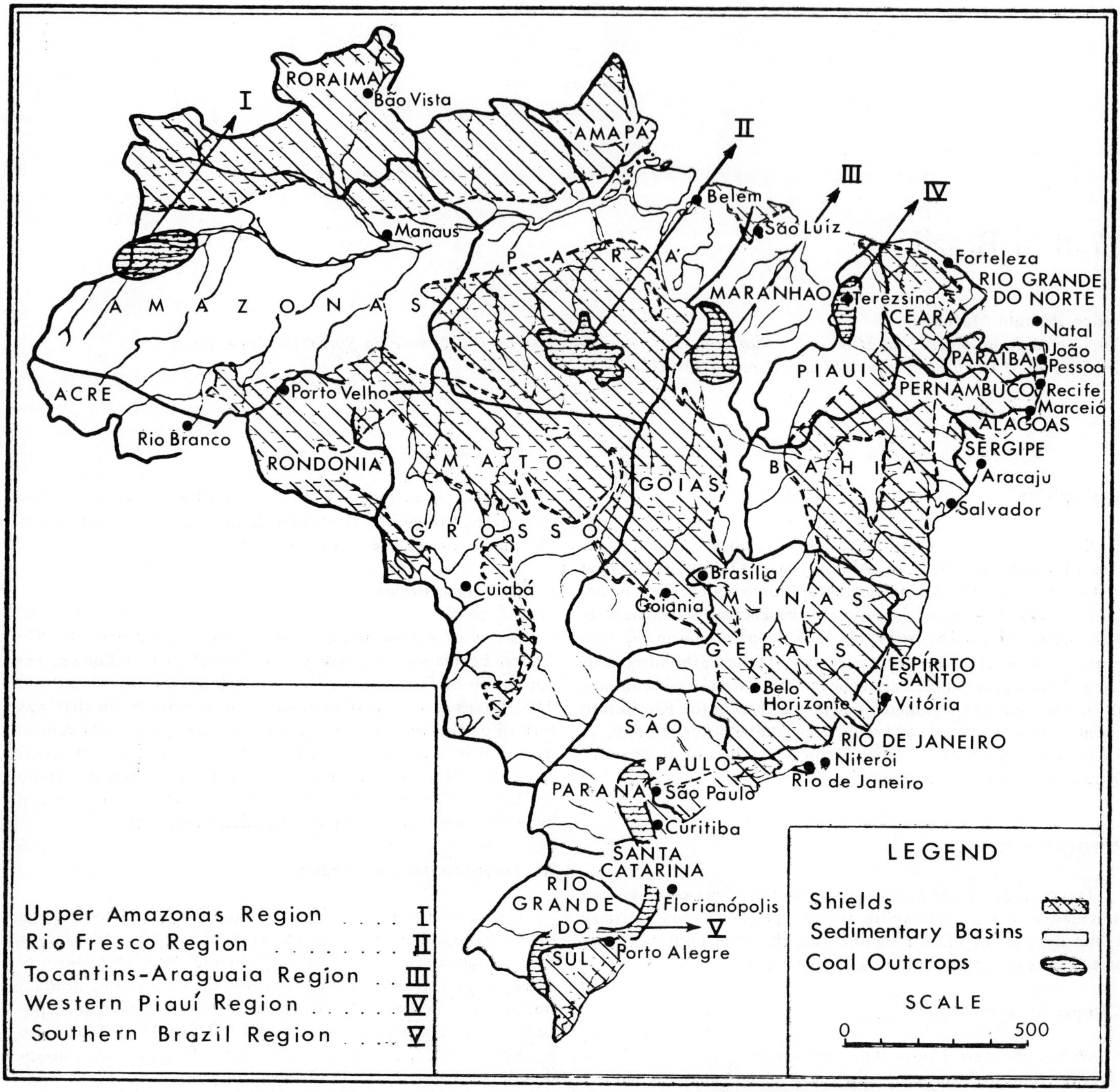

Figure 1. Index map of Brazilian coal deposits and occurrences (after Machado, 1970).

of economic deposits. CPRM and DNPM are continuing geologic studies.

V. Southern Brazil Region

In the southern Brazil region, coal occurs in a belt that extends through various areas of the states of Rio Grande do Sul, Santa Catarina, Paraná, and São Paulo. The age of the coal is controversial. The coal beds are in the basal group of the Paraná basin Gondwana sequence. This is the Tubarão Group of either Late Carboniferous or Early Permian age. There are few coal mines here, although inferred reserves are 3 billion metric tons (hereafter, metric tons = t). Only 5.6 million t are mined annually. The region is discussed below in some detail.

SOUTHERN BRAZIL REGION

Geologic Setting

The distribution of the main coal deposits of southern Brazil is shown in Figure 2. In general, the coal deposits of the Gondwana

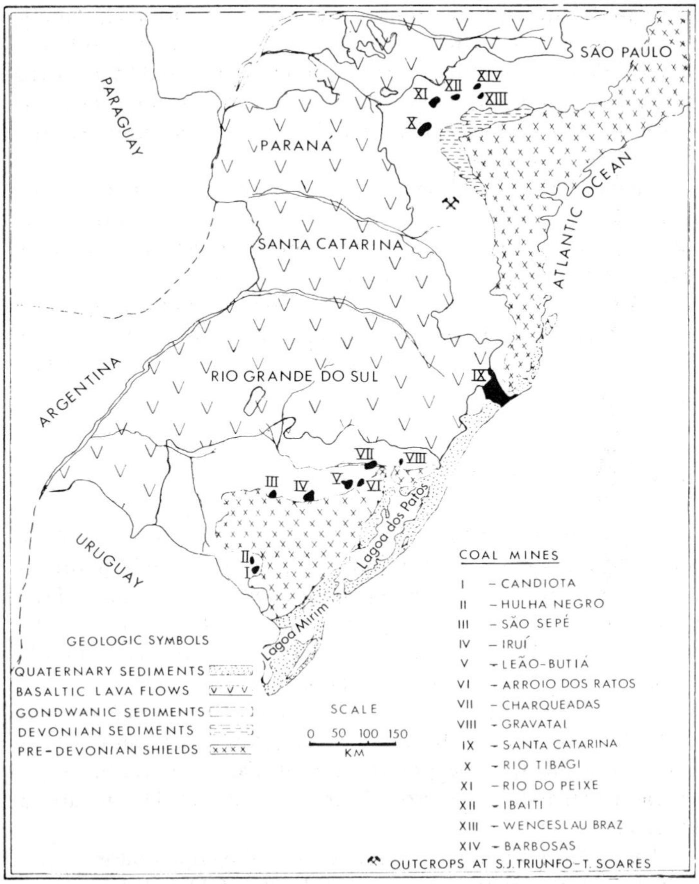

Figure 2. Geologic sketch map of southern Brazil (after Machado, 1970).

sequence crop out along the periphery of the Paraná basin (see White, 1908; Martins, 1974), between the coastal part of the Brazilian shield and the interior basalt plateau. This is a region of rugged, mountainous relief that is broken only in the zone bordering the states of Rio Grande do Sul and Santa Catarina, and by the basalt plateau. The basalt plateau is the world's largest—1,200,000 km^2—and forms a cap above the Gondwana sequence in most of the Paraná basin. The Gondwana section everywhere overlies rocks of the Brazilian shield or, locally, Devonian strata. The Gondwana section consists of the following groups (in ascending order): (1) Tubarão Group—sedimentary rocks of either Late Carboniferous or Early Permian age, (2) Passa Dois Group—Permian sedimentary rocks, and (3) São Bento Group—Triassic and Jurassic continental strata that are covered by up to 1,000 m of basalts as old as Early Cretaceous.

Tubarão Group

The coal beds are in the Tubarão Group, which is subdivided as follows (from top to bottom): (1) Guatá Subgroup (including the Palermo Formation and Rio Bonito Formation) and (2) Itararé Subgroup. In this classification, the Gondwana glacial deposits are included in the Itararé Subgroup, and the principal coal seams are included in the Rio Bonito Formation. As shown by many recent studies, however, the stratigraphic sequence of the Tubarão Group is far more complex than originally believed. The recent work has produced numerous stratigraphic controversies that are not covered in the present paper. In general, the new studies have demonstrated the existence of coal beds within the glacigene section, a change in the number and ages of tillite beds from north to south, and evidence for several limited marine transgressions in both the Itararé Subgroup and the overlying Rio Bonito Formation. These new findings have polarized those studying the area into two broad groups: (1) those who believe that the basal glacigene sedimentary deposits, whether or not they contain interbeds of coal, can be separated from the overlying Guatã Subgroup; and (2) those who believe that within the Tubarão Group as a whole, there are numerous lateral and vertical facies changes, at least along the eastern margin of the Paraná basin.

Regardless of overall interpretation, in the coal-mining region that is described here, the Tubarão Group has a fairly simple overall stratigraphy. This is true because only the postglacial coals of the Rio Bonito Formation can be economically mined. It should also be noted that, within the area described here, marine intercalations are absent in the Rio Bonito Formation. Therefore, I will describe only the formations of the Guatá Subgroup, because these are the units closely associated with the coal-mining operations of the country.

Rio Bonito Formation. This formation, whose thickness in the area normally is less than 90 m, is composed mainly of alternating layers of white to yellow-pink quartz arenite, gray to creamy shale, carbonaceous shale, and coal seams (consisting of the remains of a *Glossopteris-Gangamopteris* flora). Primary sedimentary structures indicate that the formation formed in shallow lacustrine water. These structures include current and wave ripple marks, channels, and well-developed cross-bedding. Certain features of the sandstones suggest deltaic conditions. The clay fraction is mainly kaolinitic (Formoso and Figueiredo, 1967). The upper contact is concordant with the Palermo Formation. The lower contact is with strata assigned to the Itararé Subgroup (rhythmites, diamictites, conglomerates) or, as in many areas of Rio Grande do Sul State, with rocks of the crystalline basement.

Palermo Formation

This formation (Figs. 3, 4, 5), whose thickness in the mines of the area commonly is less than 100 m, consists mainly of sandy

Figure 3. Cores of the Palermo Formation, showing the thin white bands of sandstone. Cores are from the Iruí mine, Rio Grande do Sul State.

siltstone and fine-grained silty sandstone, classified as subgraywacke. Typically, the beds are thin, undulating, or irregular with fine interlaminations of siltstone and sandstone divided by streaks of dark carbonaceous matter. The carbonaceous streaks, together with thin layers of light- and dark-gray claystone, impart a banded appearance to the section (see Figs. 4, 5). It is common to find very fine cross-laminations, some pyrite, worm borings, and delicate primary features that are characteristic of tidal flats and the inner parts of shelves. The observed primary structures are interesting to compare with similar structures reported by Shepard and Moore (1955), Shepard (1956), Wanless and others (1957), Potter and Glass (1958), and Coleman and Gagliano (1965). Previously, no conclusive evidence of a marine origin for these sedimentary rocks had been found.

The clay fraction in the Palermo Formation differs from that in the Rio Bonito Formation by being predominantly montmorillonitic (Formoso and Figueiredo, 1967). A very subordinate part of the formation is composed of thin beds and lenses of white quartz arenite, commonly very calcareous.

COAL RESERVES

Table 1 is a tabulation of coal reserves in the areas illustrated on Figure 2. I prefer not to use standard reserve classifications

Figure 4. Cores of the Palermo Formation showing other primary sedimentary structures. The cores were cut longitudinally. From Iruí mine, Rio Grande do Sul State.

Figure 5. Outcrop of the Palermo Formation, showing the thick banks of yellowish and greenish shale and resistant fine laminae of quartz arenite. Federal Highway 290, Cachoeira do Sul, Rio Grande do Sul State.

(measured, indicated, and inferred reserves), because such classifications are not applicable at present. This is true because different organizations and companies developing the coal have used different methods of estimating reserves in each mine or area. Accordingly, I have adopted the following nomenclature (see the detailed breakdown in Table 5).

Blocked Reserves. These are reserves determined from outcrops, wells, auger holes, water wells, trenches, and shafts — sufficient in number and spacing to guarantee accuracy in calculations and in coal quality. In many cases, blocked reserves correspond to measured and indicated reserves of standard classifications.

Total Estimated Reserves. This is the total reserve and includes those based on the general outcrop distribution, scattered test holes, and the known geology of the area.

MINING OPERATIONS

In Rio Grande do Sul State, mining at the present time is done on a relatively small scale at Candiota, Leão-Butiá, and Charqueadas. The Candiota mine is an open-pit operation (Fig. 6), and the principal seam has an average thickness of 4.2 m. This is the "Candiota Bed"; nine additional seams are not economically workable. The "Candiota Bed" characteristically consists of two 2-m seams separated by 90 cm of gray claystone. Although the coal is of inferior quality (Table 2), the open-pit mining keeps the costs low. Consequently, the powerhouse at Candiota produces the cheapest thermally generated electricity in Brazil. Details are in Machado (1975).

The Leão-Butiá mine, originally worked by open-pit methods, now is an underground operation at depths of 30 to 127 m. There are two seams, S_1 and S_2. The two are separated by 2 to 15 m of shale. The lower S_2 bed is the more important and has an average

TABLE 1. COAL RESERVES IN SOUTHERN BRAZIL

Deposit	Blocked	Total	Type
	Rio Grande do Sul State		
Candiota (including Hulha Negra)	1,265.0	8,000.0	Subbituminous
São Sepé	3.5	7.0	Subbituminous
Iruí	109.0	330.0	Subbituminous
Leão-Butiá	40.0	640.0	Subbituminous
Arroio dos Ratos	Mined	Mined	Subbituminous
Charqueadas	1,013.0	1,100.0	Subbituminous
Gravataí	12.0	15.0	Subbituminous
Subtotal	2,442.5	10,092.0	
	Santa Catarina State		
Santa Catarina	760.0	1,500.0	Bituminous
	Paraná State		
Rio Tibagi	5.8	7.0	Bituminous
Rio do Peixe	29.0	40.0	Bituminous
Ibaití	0.2	1.4	Bituminous
Wenceslau Braz	0.0	0.1	Bituminous
Barbosas	0.2	0.6	Bituminous
Subtotal	35.2	49.1	
Total	3,237.7	11,641.1	

Note: Reserves in millions of metric tons. Blocked=blocked reserves as defined in text; total=total estimated reserves as defined in text. Data sources are reports of DACM, CRM, COPELMI, DNPM, and CPRM.

thickness of 1.3 m. This coal is the best in Rio Grande do Sul (Table 2). Details are given by Machado (1975).

The Charqueadas mine is mined by underground shafts and tunnels at a depth of about 270 m. There are five seams, of which only one is worked (Table 2). The main seam is in three layers — one about 40 cm thick, a principal layer about 1 to 2 m thick (average is 1.4 m); and a lower bed 10 to 30 cm thick. Details are given in Machado (1975).

Two mining organizations operate in Rio Grande do Sul. One belongs to the state (Sociedade de Economia Mista).

In Santa Catarina State, the Barro Branco seam is mined intensively (Fig. 7). Another seam, the Irapuá, is mined locally and on a much less intensive scale. The total bed thickness of the Barro Branco seam ranges from 1.4 to 2.2 m; the thickness of recoverable coal ranges from 0.5 to 1.0 m. Coal characteristics are shown in Table 3. The Irapuá seam ranges in thickness from zero (a carbonaceous shale) to 1.12 m. Both are mined by open-pit and underground methods. The underground mines are inclined shafts, with inclinations of 3° to 18°. The maximum depth of operations is less than 140 m.

Thirteen organizations are developing the coal. The largest of these has an organization similar to the state-owned Sociedade de Economia Mista in Rio Grande do Sul. At present, in accord with government policy, the large companies are merging to establish a smaller number of mining consortiums that can operate more economically.

In Paraná State, two private mining companies operate. One mines the Rio de Peixe area, and the other operates the Rio Tibagi area (Fig. 2). In both areas, there is only one economically mineable coal seam whose average thickness is less than 1 m. Depths to the seam range up to 130 m. Coal characteristics are shown in Table 4.

CHARACTERISTICS OF COAL BEDS

The coal seams discussed here occur in gently dipping, asymmetrical synclines or in structural basins. The type of structure suggests that the area never underwent diastrophism; instead, the structure suggests that compaction processes produced the gentle synclines, with the least amount of compaction around the margins of the structural basins where the volume of clastic rocks is greatest (Machado, 1960). Thus the tectonic regime of the region is one of epeirogenesis. As a result of these essentially vertical movements, linear faulting took place under tensile-stress conditions (Putzer, 1952, 1953).

In terms of quality, the southern Brazil coals generally are inferior. This, in my opinion, is the result of two factors: (1) the coal beds formed in a rather cold climate (as inferred from the *Glossopteris-Gangamopteris* flora) in which the woody tissue developed specialized characteristics of fibers, resin, and so forth and (2) the coal beds formed in an area of little tectonism, dominated by epeirogenic movements. The fundamental coal types are bituminous and subbituminous (Table 1). Details of the coal lithology are not given here, but will be published in subsequent papers (see also Machado, 1975).

Figure 6. The Candiota seam at Candiota mine, Rio Grande do Sul State. The seam here consists of two beds of coal, each about 2.0 m thick and separated from one another by 70 cm of light-gray shale.

Figure 7. Typical appearance of the Barro Branco seam, Siderópolis mine, Santa Catarina State.

TABLE 2. TYPICAL PROXIMATE ANALYSES (DRY BASIS) OF COAL FROM RIO GRANDE DO SUL MINES

	Candiota	Iruí	Leão-Butiá	Charqueadas
Moisture (%)	11.30	9.60	5.00	6.70
Volatile matter (%)	21.00	24.30	25.60	19.50
Fixed carbon (%)	26.60	33.80	34.80	28.90
Ash (%)	52.40	41.80	39.60	53.60
Sulfur (%)	2.20	0.30	0.70	0.70
Highest calorific value (cal/g)	3,248	4,158	4,323	3,151

TABLE 3. TYPICAL PROXIMATE ANALYSES (DRY BASIS) OF ZONES IN BARRO BRANCO SEAM, SANTA CATARINA

	Metropolitana	Criciuma	Próspera	Urussanga
Moisture (%)	2.70	3.20	2.50	3.00
Volatile matter (%)	27.10	25.20	24.80	25.50
Fixed carbon (%)	40.70	39.30	40.90	33.80
Ash (%)	29.50	32.30	31.80	37.70
Sulfur (%)	3.60	5.50	4.50	5.10
Highest calorific value (cal/g)	5,575	5,000	5,220	5,750

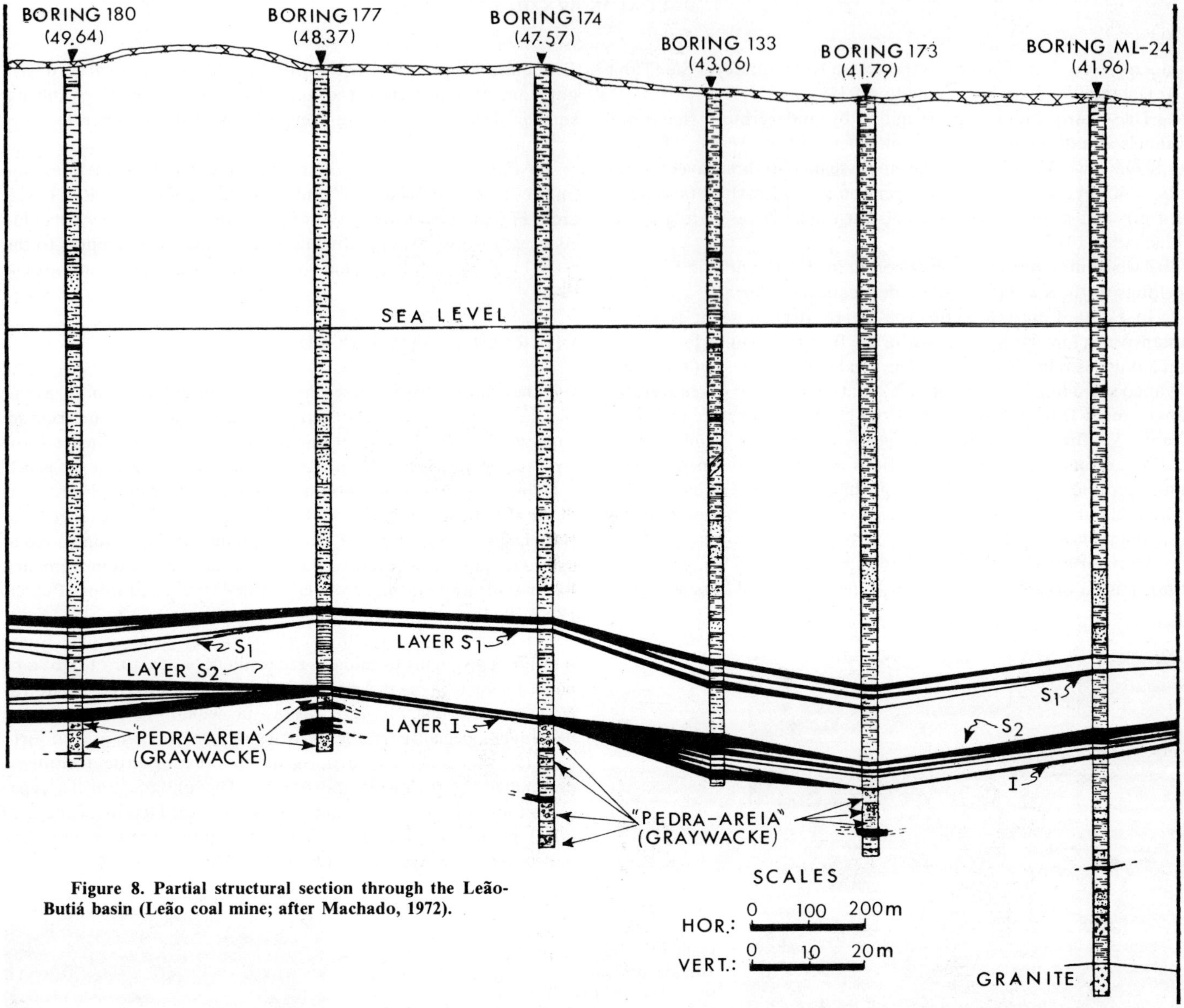

Figure 8. Partial structural section through the Leão-Butiá basin (Leão coal mine; after Machado, 1972).

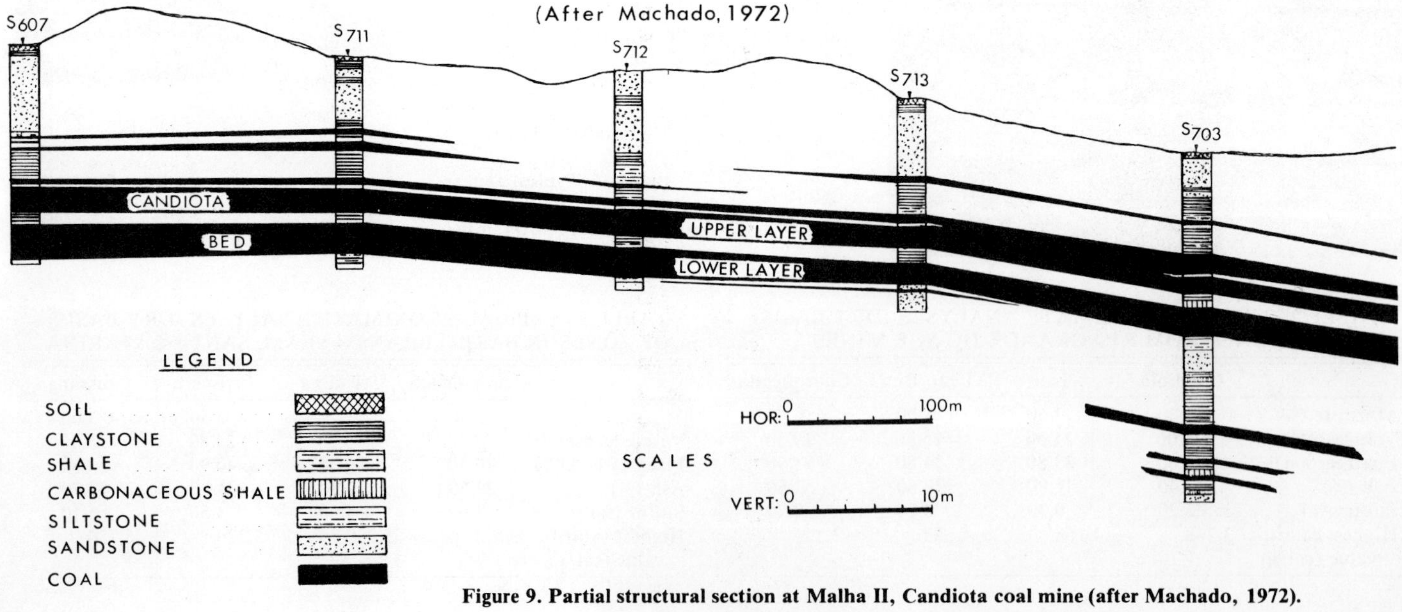

Figure 9. Partial structural section at Malha II, Candiota coal mine (after Machado, 1972).

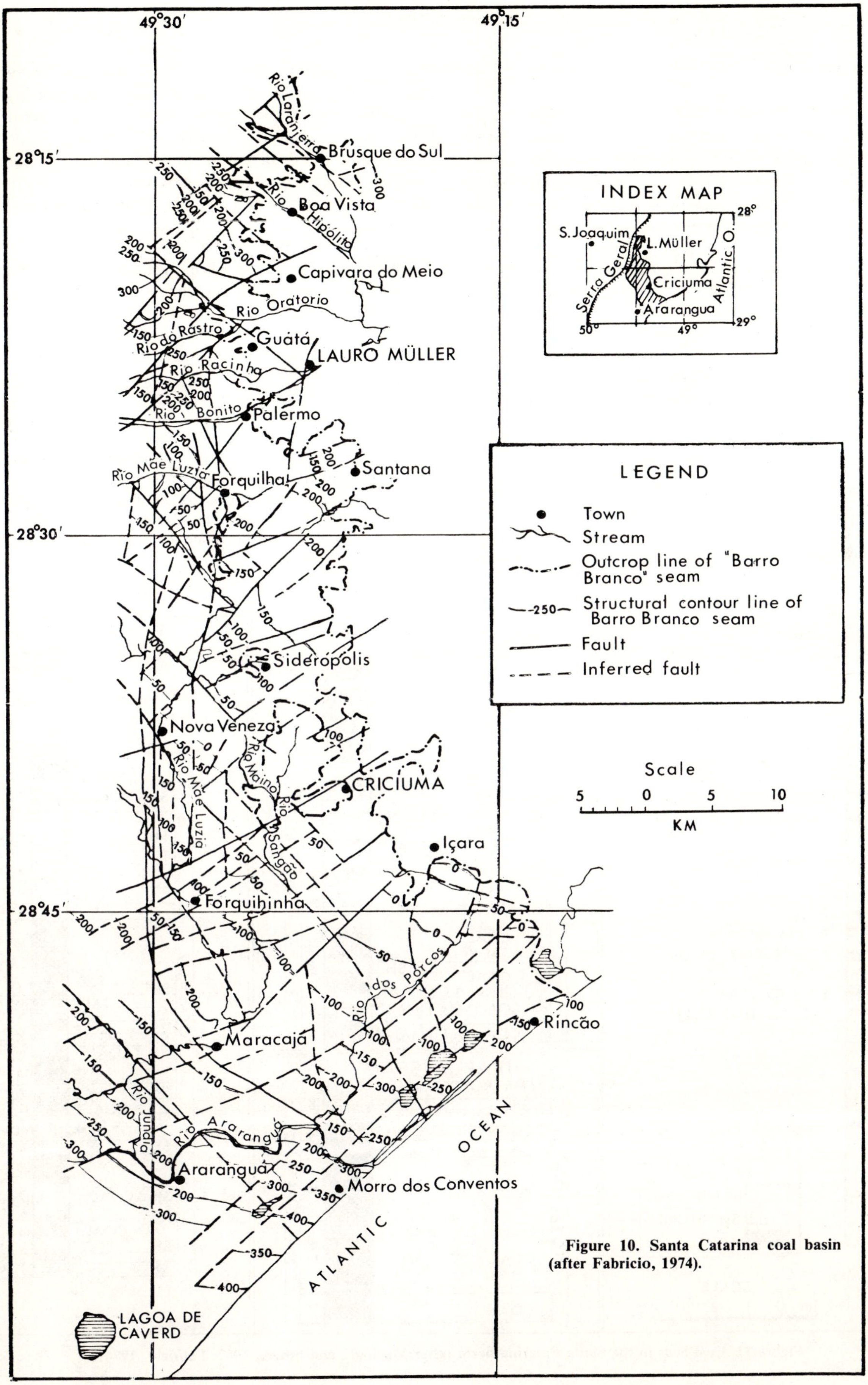

Figure 10. Santa Catarina coal basin (after Fabricio, 1974).

Rio Grande do Sul

Several coal seams are present in Rio Grande do Sul State, but because of their properties, only one or two are mineable. Thicknesses normally are less than 2 m, with the exception of the "Candiota Bed" which has a thickness of as much as 4.2 m. Figures 8 and 9 give partial sections of the Candiota and Leão-Butiá areas and illustrate the typical distribution and structure of the coal mines of the state. Table 2 shows dry-base analyses of Rio Grande do Sul coals before beneficiation. The samples from Iruí and Leão-Butiá were hand-selected from jigs. The coals are noncoking types.

Santa Catarina

The coal occurrences of this state form a single deposit. It is the only coal in Brazil that is capable of producing a coking coal for steel manufacture. Figure 10 shows that the coal-bearing area of Santa Catarina is oriented north-south. It is 80 km long and 20 km wide, with an area of approximately 1,050 km^2.

Between March 1971 and October 1973, two Federal agencies (DNPM and CPRM) began an intensive exploration program with the objective of determining the position of Santa Catarina coals in the world coal market. It is evident that the geologic data obtained by this exploration effort will be extremely useful in evaluating the ultimate potential of the Santa Catarina coal-bearing area (Fabricio, 1974; Willig, 1974). As a result of the exploratory effort, the following important facts emerged.

1. Test holes drilled in the southern part of the Santa Catarina coal district revealed several new coal seams, previously undescribed in the literature (Fig. 11). From top to bottom, the coal seams were named Treviso, Barro Branco, Irapuá, A, B, Ponte Alta, Bonito, Pré-Bonito, C, and D (see Figs. 11, 12). Of these, the Barro Branco, Irapuá, and Bonito seams are mineable (at present, the Bonito seam is not being mined).

2. With regard to the structure of the Barro Branco seam, the most important in Santa Catarina, Fabricio's (1974) conclusions were substantiated (Fig. 12). Specifically he wrote,

> The basin has the structure of a large monocline with a regional N70°W strike, a gentle regional plunge, less than 1° toward the southwest, and a great density of faults. The strike of the seam, as well as of the whole stratigraphic column, changes gradually toward the east in the southern part of the basin. Local changes in plunge, commonly in opposite directions, and the existence of locally developed anticlines and synclines, are the results of movements along the pre-existing faults and of igneous intrusions (diabases).

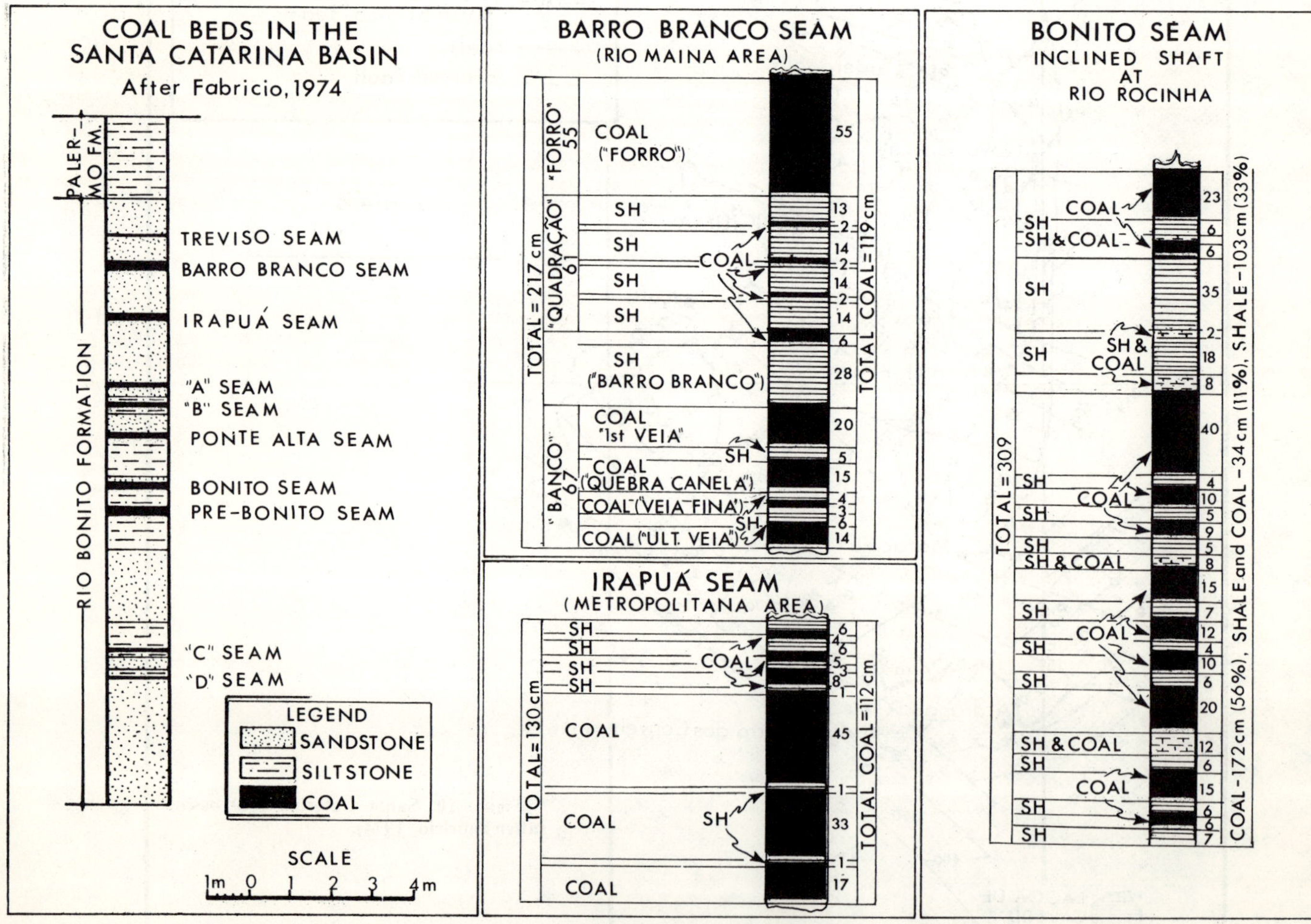

Figure 11. Coal beds in the Santa Catarina basin (after Machado and others, 1962; Fabricio, 1974).

The zero elevation line of the Barro Branco seam in general extends from the mouth of the Rio dos Porcos to Nova Veneza. North and northeast of this line, the seam rises as much as 350 m, and south and southwest of the line, the seam reaches depths as great as 350 m.

The numerous faults in the area offset the coal seams in many places. The faults are of the normal type, and have three strike directions. The result is the creation of a mosaic pattern, locally with vertical plunges.

The three fault directions are northeast, north, and northwest.

The northeast fault direction is the principal one, and causes major problems in mining, because it cuts all of the principal concentrations of coal. Most of these faults are 30 to 40 km long, with throws that range from 30 to 110 m.

The north-south faults generally are long, up to 50 km, with throws up to 100 m. These faults are present mainly in the eastern margin of the coal-bearing area; because of this, they rarely affect mining operations adversely.

The last, or northwest, fault system rarely has faults longer than 10 km or throws greater than 20 m. They are most common in the area northeast of the town of Lauro Muller.

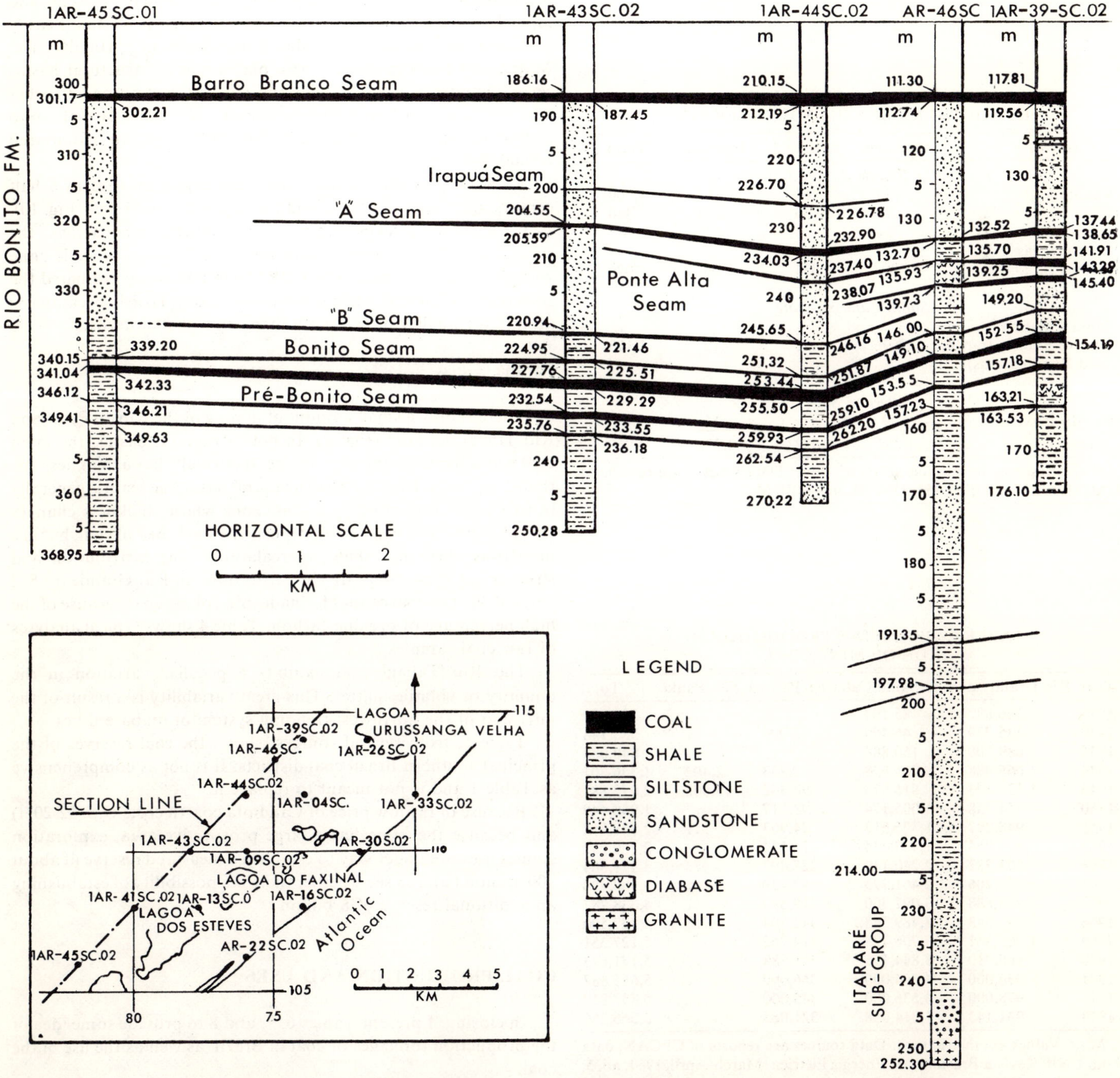

Figure 12. Partial structural section through Santa Catarina coal basin (modified from Fabricio, 1974).

TABLE 4. TYPICAL PROXIMATE ANALYSES (DRY BASIS) OF TWO MINE AREAS, PARANA

	Rio do Peixe			Rio Tibagi		
Moisture (%)	2.36	4.40	3.00	2.00	1.90	4.50
Volatile matter (%)	29.00	32.90	22.40	10.00	20.80	24.70
Fixed carbon (%)	46.50	40.80	51.10	51.40	41.00	36.30
Ash (%)	22.14	21.90	23.50	36.60	36.30	34.50
Sulfur (%)	10.00	8.22	6.63	13.69	10.60	10.40
Highest calorific value (cal/g)	5,626	5,750	5,598	4,646	4,848	4,230

TABLE 5. RESERVES OF THE PRINCIPAL MINES OF SOUTHERN BRAZIL, NOVEMBER 1974

Coal Mines	Measured	Indicated	Inferred	Total
Rio Grande do Sul State				
Candiota	175	65	1,460	1,700
Iruí	70	36	224	330
Leão-Butiá	34	6	600	640
Charqueadas (Seams I_{1F} and II_{2B})	680	333	63	1,076
Santa Catarina State				
Santa Catarina (Barro Branco, Irapuá, and Bonito seams)	269	492	412	1,173
Paraná State				
Rio do Peixe	29	..	11	40
Total	1,257	932	2,770	4,959

Note: Reserves in millions of metric tons. Data sources are reports of DACM, CRM, COPELMI, DNPM, and CPRM.

TABLE 6. BRAZILIAN PRODUCTION OF RUN-OF-MINE COAL

Year	Rio Grande do Sul	Santa Catarina	Paraná	São Paulo	Total
1925	306,682	85,197	..	..	391,879
1930	335,739	46,409	3,000	..	385,148
1935	689,200	150,888	..	..	840,088
1940	1,065,488	265,638	2,733	2,402	1,336,301
1945	1,139,853	815,678	98,342	19,002	2,072,881
1950	854,758	1,005,174	98,717	..	1,958,649
1955	948,297	1,325,512	74,903	..	2,348,712
1960	646,264	1,439,434	74,534	..	2,160,232
1965	903,732	2,240,140	227,492	..	3,371,364
1966	731,206	2,461,775	187,439	..	3,380,420
1967	925,888	3,097,300	315,599	..	4,338,787
1968	995,543	3,489,543	342,504	..	4,827,590
1969	1,005,861	3,706,728	414,762	..	5,127,351
1970	965,010	3,844,775	361,888	..	5,171,673
1971	956,000	4,364,000	346,000	..	5,665,887
1972	978,000	4,536,000	345,000	..	5,857,839
1973	931,143	4,314,058	321,065	..	5,566,266

Note: Values are metric tons. Data sources are reports of CPCAN, data from CNP, Revista Brasileira de Energia Elétrica (March-April, 1964, no. 5, Ed. da Electrobrás), and IBGE.

According to Machado and others (1962), data collected by V. Dequech (in Machado and others, 1962) show that the Barro Branco bed ranges in thickness from 1.4 to 2.2 m, with recoverable coal thicknesses ranging from 0.5 to 1.2 m. Table 3 shows some dry-base analyses of the Barro Branco seam in four localities. The coals were taken from the preliminary washers, but had scarcely been washed.

Finally, the CPRM-DNPM study (made in 1974) evaluated the total reserve of coal in Santa Catarina State (Barro Branco seam) as 760,400,000 t.

3. The Irapuá seam, second in importance within the state, is a bituminous coal with coking properties. Where present and mineable, it is a thick seam consisting of a single layer and is easily worked (Fig. 11). It is workable in two areas in particular: Rio Maina and Siderópolis. In these places it is in structural basins several hundreds of meters wide and a few kilometers long. In other areas, it is only a few centimeters thick and grades laterally into carbonaceous shale. The CPRM-DNPM group calculated 10,000,000 t of reserves.

4. Mining of the Bonito seam were begun more than a half century ago, particularly around the city of Lauro Muller (Fig. 10). Here the seam is thickest and more mineable than other areas of the state. Laterally the seam grades into a carbonaceous shale zone with thin layers of coal. The CPRM-DNPM study evaluated the reserves at 402,400,000 t. It is not mined today, largely because it is an inferior quality of coal with many thin shale interbeds.

Paraná

There is only a small amount of mining of the Rio do Peixe and Rio Tibagi deposits (Fig. 2). In both structural basins, the main coal bed is barely economic to mine. It normally has a thickness less than 1 m. In the Rio do Peixe structural basin, the seam is generally in two layers separated by a shale zone whose thickness changes considerably. The Rio Tibagi deposit is a single bed of coal, but has numerous very thin shale intercalations. The distribution and structure of these coals are similar to those in Rio Grande do Sul (Figs. 8, 9). The coal cannot be made into coking coal because of the high percentage of organic carbon. Table 4 shows typical analyses of raw coal samples.

The Rio Tabagi coal exhibits a peculiar variation in the quantity of volatile matter. This great variability is a result of the intrusion of the northwest-trending system of diabase dikes.

Table 5, from several sources, shows the coal reserves of the principal southern Brazil coal districts. It is not as comprehensive as Table 1 and is not meant to replace it.

Because of the low price of Candiota coal (it costs U.S. $2.20/t) and because the mine has a large prospective area, exploration studies now are under way to establish a measured reserve of about 900 million t and to see whether there is a possibility of establishing an additional reserve of 8 billion t.

COAL PRODUCTION AND USES

In closing, I present Tables 6, 7, and 8 to provide some idea of the production tonnages of coal in Brazil, as well as the use of the coal.

Two important facts should be noted. In 1970, the company

TABLE 7. CONSUMPTION OF COAL IN BRAZIL

	National steam coal		National metallurgical coal		Foreign metallurgical coal	
	1960	1973	1960	1973	1960	1973
Thermoelectricity (%)	55.37	93.70	..	..	..	..
Metallurgy (%)	..	..	86.97	99.30	66.36	98.40
Railroads (%)	41.29	0.30	..	..	..	..
Navigation (%)	2.86	..	..	..	..	..
Gas production (%)	..	..	13.03	0.70	33.64	0.60
Industry—various uses (%)	2.48	6.00	..	..	..	..
Total (metric tons)	821,564	1,576,834	455,193	792,697	928,296	1,698,000

Note: Data sources from CPCAN, CNP, and Revista Carvão da Pedra.

TABLE 8. COAL-FIRED POWER PLANTS IN SOUTHERN BRAZIL

Power plant and state	Capacity (MW) Actual in 1974	Proposed capacity to 1981
Candiota, Rio Grande do Sul	126	426
Charqueadas, Rio Grande do Sul	72	72
São Jerônimo, Rio Grande do Sul	20	20
Capivari (CSN), Santa Catarina	24	24
Sotelca, Santa Catarina	232	482
Figueira, Paraná	30	155
Pres. Vargas (Klabin), Paraná	27	27

Note: Data sources from CPCAN and various other organizations.

Acos Finos Piratiní S.A. in Rio Grande do Sul State, close to the Charqueadas mine, began operations to produce special steel by the SL/RN process of direct reduction. Furthermore, in Ibituba, Santa Catarina State, the Industria Carboquímica Catarinense S.A. began to utilize pyritic wastes from the washing of the Santa Catarina coal to produce sulfuric acid and phosphoric acid. The Rio Grande do Sul coals were tested in West Germany at the Lurgi and Koppers plants, as well as in the Institute of Coal Investigations in Essen.

The results obtained were excellent and proved that the coal can be used for steel manufacture (on the basis of the direct reduction process) and for the production of gas. For this reason, viability studies now are in progress to install large plants that will utilize these techniques.

REFERENCES CITED

Coleman, J. M., and Gagliano, S. M., 1965, Sedimentary structures—Mississippi River deltaic plain, *in* Primary sedimentary structures and their hydrodynamic interpretation: Soc. Econ. Paleontologists and Mineralogists Spec. Pub. 12, p. 133–148.

Fabricio, J.A.C., 1974, A jazida de carvão de Santa Catarina: Porto Alegre, Companhia de Pesquisa de Recursos Minerais, 21 p., 8 maps (unpub.).

Formoso, M. L., and Figueiredo, P. M., 1967, Estudo sedimentológico e petrológico do Grupo Guatá: Univ. Fed. Rio Grande do Sul, Escola Geologia, Pub. Esp. no. 12, 66 p.

Machado, E. R., 1960, Contribução ao e studo do carvão mineral no Rio Grande do Sul: Univ. Fed. Rio Grande do Sul, Escola Geologia, Pub. Esp. no. 3, 31 p.

——1970, Geologia do carvão no Brasil, *in* III Simposio do carvão nacional, Curitiba: Rev. Carvão de Pedra, nos. 11/12, ano 2, p. 26–39.

——1972, O carvão Neopaleozóico do Brasil meridional: Acad. Brasileira Ciénc. Anais, v. 44 (supplement), p. 209–235.

——1975, Gondwana coal in southern Brazil, in Campbell, K.S.W., ed., Gondwana geology (Papers presented at the Third Gondwana Symposium Canberra, Australia, 1973): Canberra, Australian Natl. Univ. Press, p. 271–293.

Machado, E. R., Dequech, V., and Castanho, O. S., 1962, Condicionamento geológico das jazidas de carvão do Brasil meridional: Univ. Fed. Rio Grande do Sul, Escola Geologia, Avulso no. 4, 55 p.

Martins, A.J.M., 1974, Carvão no Paraná: Porto Alegre, Companhia de Pesquisa de Recursos Minerais, 11 p. (unpub.).

Potter, P. E., and Glass, H. J., 1958, Petrology and sedimentation of the Pennsylvanian sediments in southern Illinois—a vertical profile: Illinois Geol. Survey Rept. Inv. 204, 60 p.

Putzer, H., 1952, Camadas de carvão mineral e seu comportamento em Santa Catarina: Dept. Nac. Prod. Min., Bol. 91, 182 p.

——1953, Diastrofismo germanótipo e sua relacão com o vulcanismo basáltico na parte meridional de Santa Catarina: Soc. Brasileira Geologia Bol., v. 2, no. 1, p. 37–74.

Shepard, F. P., 1956, Marginal sediments of Mississippi delta: Am. Assoc. Petroleum Geologists Bull., v. 40, p. 2537–2623.

Shepard, F. P., and Moore, D. G., 1955, Central Texas coastal sedimentation—characteristics of sedimentary environment, recent history, and diagenesis: Am. Assoc. Petroleum Geologists Bull., v. 39, p. 1463–1593.

Wanless, H. R., and others, 1957, Limestone texture as a key to interpreting depth of deposition, *in* Relaciones entre la tectónica y la sedimentación, t. 1: Internat. Geol. Cong., 20th, México 1956, sec. 5, p. 65–82.

White, I. C., 1908, Relatório sobre as coal measures e rochas asociadas de sul do Brasil: Com. Est. Minas Carvão Pedra Brasil, Relatório Final, pt. 1, 300 p.

Willig, C., 1974, Trabalhos de pesquisa de carvão realizados pelo convénio DNPM/CPRM em Santa Catarina: Conselho Nacional Petróleo Atual., no. 38, ano 5, p. 8–55.

MANUSCRIPT RECEIVED BY THE SOCIETY OCTOBER 27, 1976
MANUSCRIPT ACCEPTED DECEMBER 21, 1977

Printed in U.S.A.

Geological Society of America
Special Paper 179

Coal resources of Peru

C. Richard Petersen*
Administração e Comércio, Ltda., Caixa Postal 5052-ZC37, Rio de Janeiro GB, Brazil

ABSTRACT

The coal deposits of Peru are in three facies of the Lower Cretaceous Goyllarisquizga Group—the eastern, southern, and western facies. Only the western facies is divided into formations, which are from bottom to top the Oyón, Chimú, Santa, Carhuaz, and Farrat. The most important coalfields are the Goyllarisquizga, Jatunhuasi, Oyón, and the "Northern Anthracite" fields.

The Goyllarisquizga field, in constant production since 1903, has yielded at least 8,800,000 metric tons of bituminous coal, which has been used to make metallurgical coke for the Cerro de Pasco Corporation's smelter in La Oroya. The coal produced in 1967 averaged 54.0 percent ash, 21.5 percent volatile matter, and 21.5 percent fixed carbon.

The Jatunhuasi field is a long, narrow body of coal east of the continental divide. It has been explored by the Cerro de Pasco Corporation—now Empresa Minera del Centro del Perú (Centromin)—because it constitutes the company's main source of supply with a resource potential of about 8 million metric tons, although proved reserves are considerably less. The coal beds exhibit characteristics different from those in the Goyllarisquizga field, and the ash content is in general lower than that for coal in the Goyllarisquizga field.

In the Oyón basin both anthracite and bituminous coals occur in the Oyón Formation. Variations in rank of the coals are either the result of intrusion of granodiorite stocks or of intense tectonic activity during which most of the coal was sheared and shattered.

The coal in northern Peru is of anthracitic rank. The coal-bearing rocks in this area extend from the Santa River valley to the upper Chicama River. Coal beds occur in both the Chimú and Farrat Formations. By far the most important area is the region of the upper Chicama, where a substantial tonnage could be developed if a suitable market could be found.

INTRODUCTION

This paper describes the general geologic setting of the most important coalfields in Peru and gives an idea of the potential of these fields based on the latest information available, which is contained principally in unpublished private reports. The literature on coal in Peru is extensive, but most articles have incomplete descriptions of isolated occurrences, which have no possibility of becoming a commercial mine. Some older publications have a tendency to exaggerate the resource potential of some of the important coalfields.

Most of the information for this paper has been made available through the cooperation of the Cerro de Pasco Corporation, the Servicio de Geología y Minería of the Peruvian Government, and the Banco Minero del Peru.

GEOLOGIC SETTING

The principal coalfields in Peru are in the central and northwestern parts of the country (Fig. 1). They are of Early Cretaceous age and are contained within the sandstone and quartzite of the so-called Goyllarisquizga Group. The position of this group within the Peruvian stratigraphic sequence is illustrated in Figure 2. The names of most of the units are used mainly in central Peru, but the Goyllarisquizga Group can be correlated in most parts of the country.

Wilson (1963) recognized three distinct facies within the Goyllarisquizga Group and called these the eastern, southern, and western facies (see diagrammatic section, Fig. 3). The eastern facies consists predominantly of orthoquartzite located mainly east of the continental divide and northwest of Cerro de Pasco in central Peru. The facies is regarded as being of Neocomian to earliest Albian age (Fig. 3), because it is overlain by deposits of Albian age (Pariahuanca and Chulec Formations). The eastern facies probably was deposited in deltas or on river floodplains. The Goyllarisquizga coalfield, which has produced a considerable amount of coal, is within this facies.

The southern facies extends south of the Junín area as far as the Huancavelica region (Fig. 1). It consists of variable amounts of sandstone and shale and contains the Jatunhuasi coalfield.

The western facies is much thicker and more varied lithologically than its eastern and southern equivalents. It extends from the Canta region northwestward into the northern Andes, west of the continental divide. This facies of the Goyllarisquizga Group has been subdivided into the following formations in ascending order:

*Present address: U.S. Steel Western Hemisphere, Inc., Suite 1208, 7 King Street East, Toronto, Ontario, Canada.

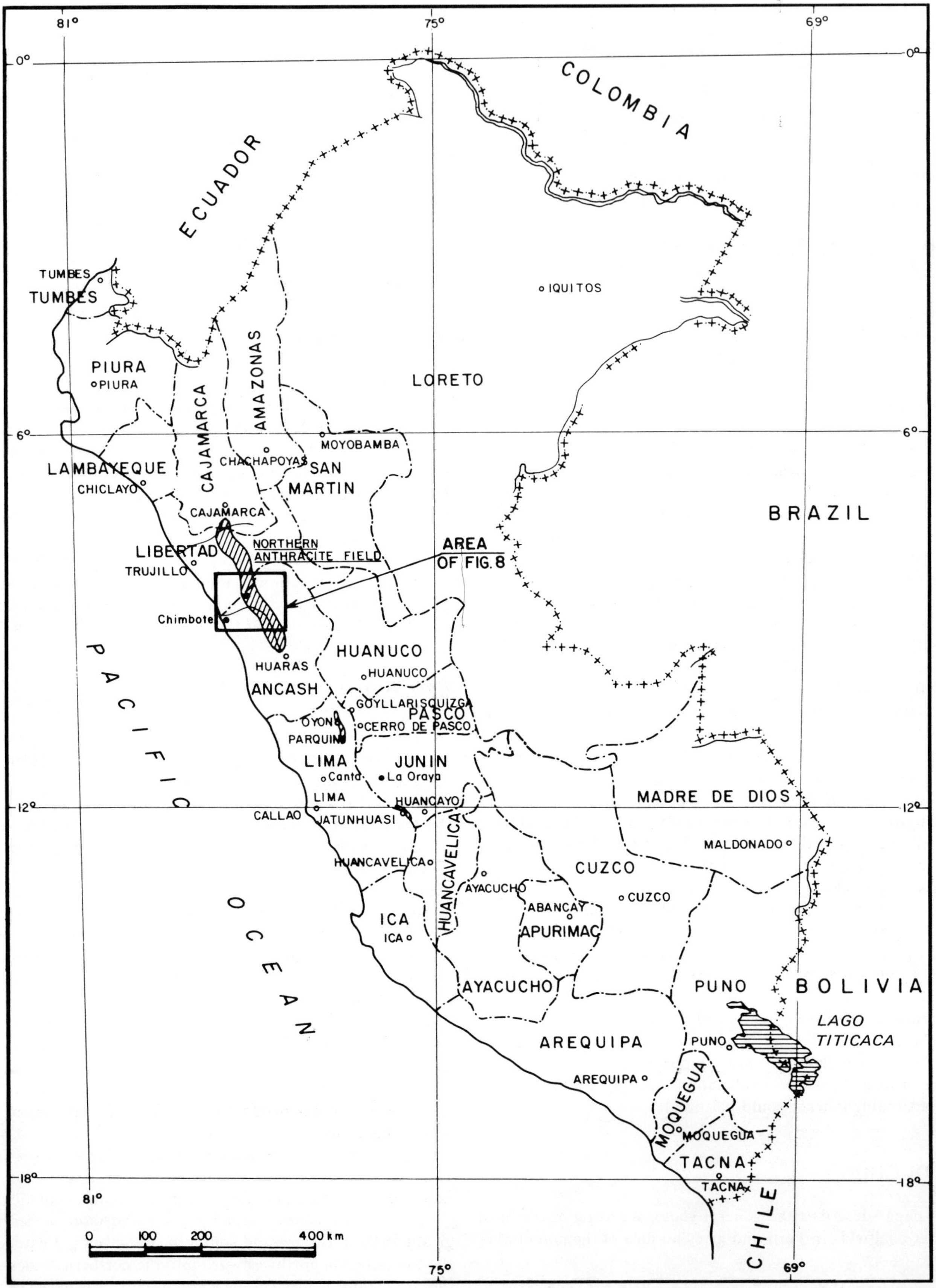

Figure 1. Location map of the principal coalfields in Peru.

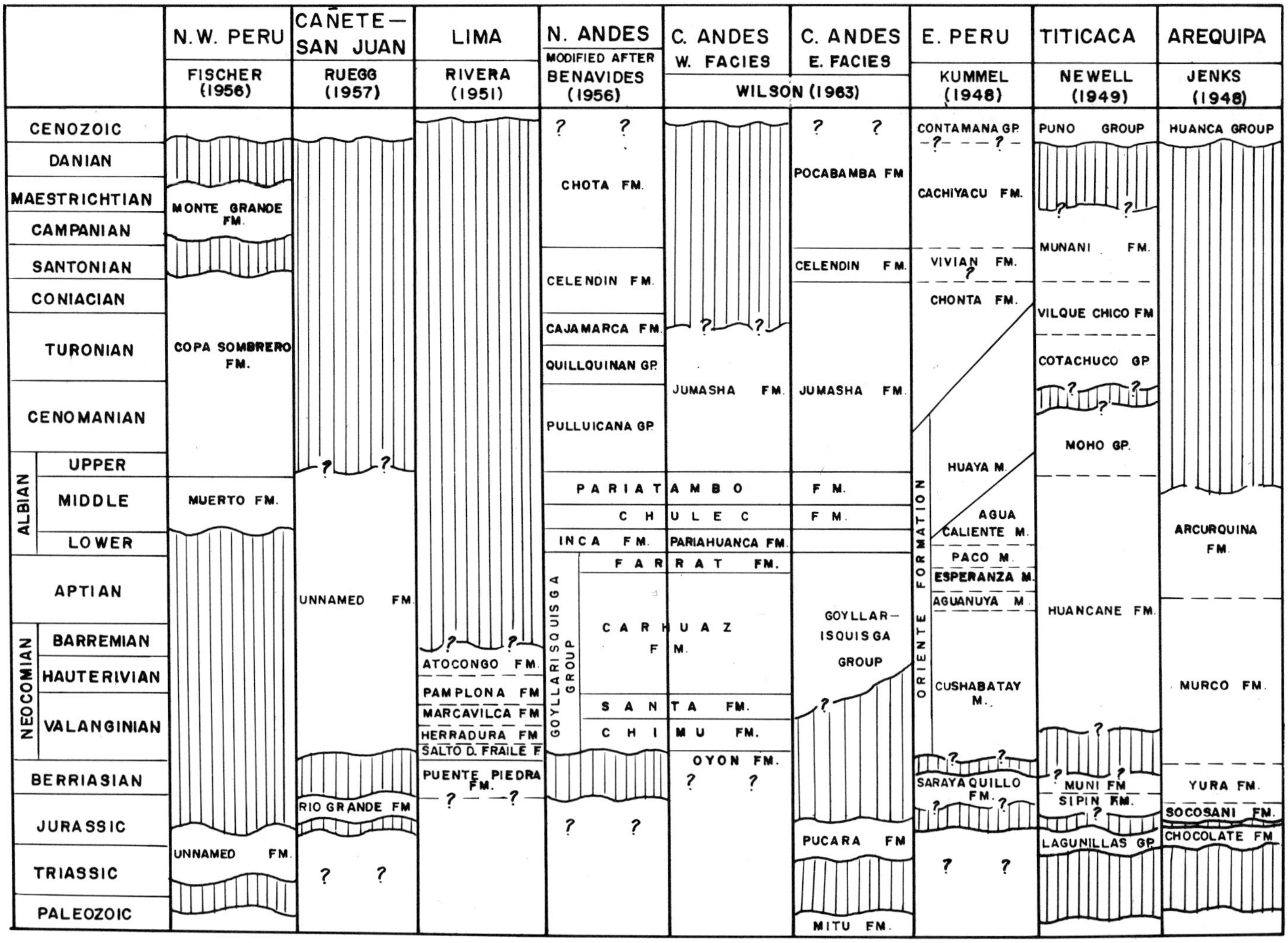

Figure 2. Correlation table for the Peruvian stratigraphic column (after Wilson, 1963).

Oyón, Chimú, Santa, Carhuaz, and Farrat Formations, all of which contain coal in the sandstone beds, except the Santa Formation.

COALFIELDS

Goyllarisquizga Coalfield

Location. The Goyllarisquizga coalfield is about 30 km north of Cerro de Pasco City in central Peru (Fig. 1). The coal beds are in the eastern facies of the Goyllarisquizga Group. The coals are probably Neocomian, but a more accurate age determination has not been made.

Geology. In the Goyllarisquizga field, a close relationship exists between stratigraphy and tectonic structure (Figs. 4, 5). A complex southeast-plunging syncline exhibits thickening of the individual stratigraphic units toward the synclinal axis. The coal beds tend, therefore, to thin out on the flanks of the synclinal structure. This field has yielded nearly 8,800,000 metric tons of coal since 1903, which is the earliest year for which production records are available. The mine was closed at the end of 1971 for economic reasons.

The Goyllarisquizga Group (Figs. 4, 5) has been subdivided

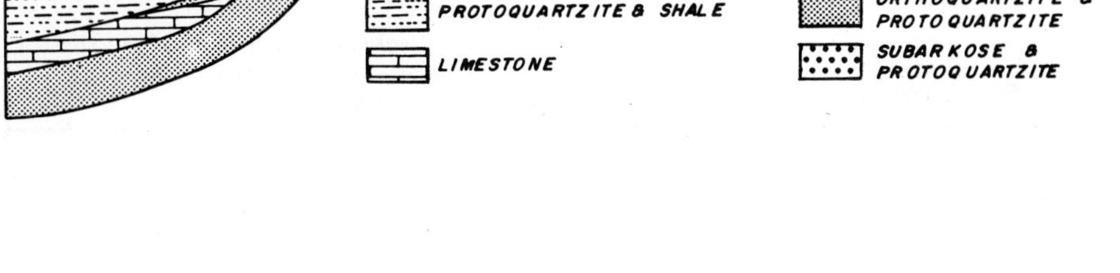

Figure 3. Lithologic relationships between the facies of Goyllarisquizga Group (after Wilson, 1963).

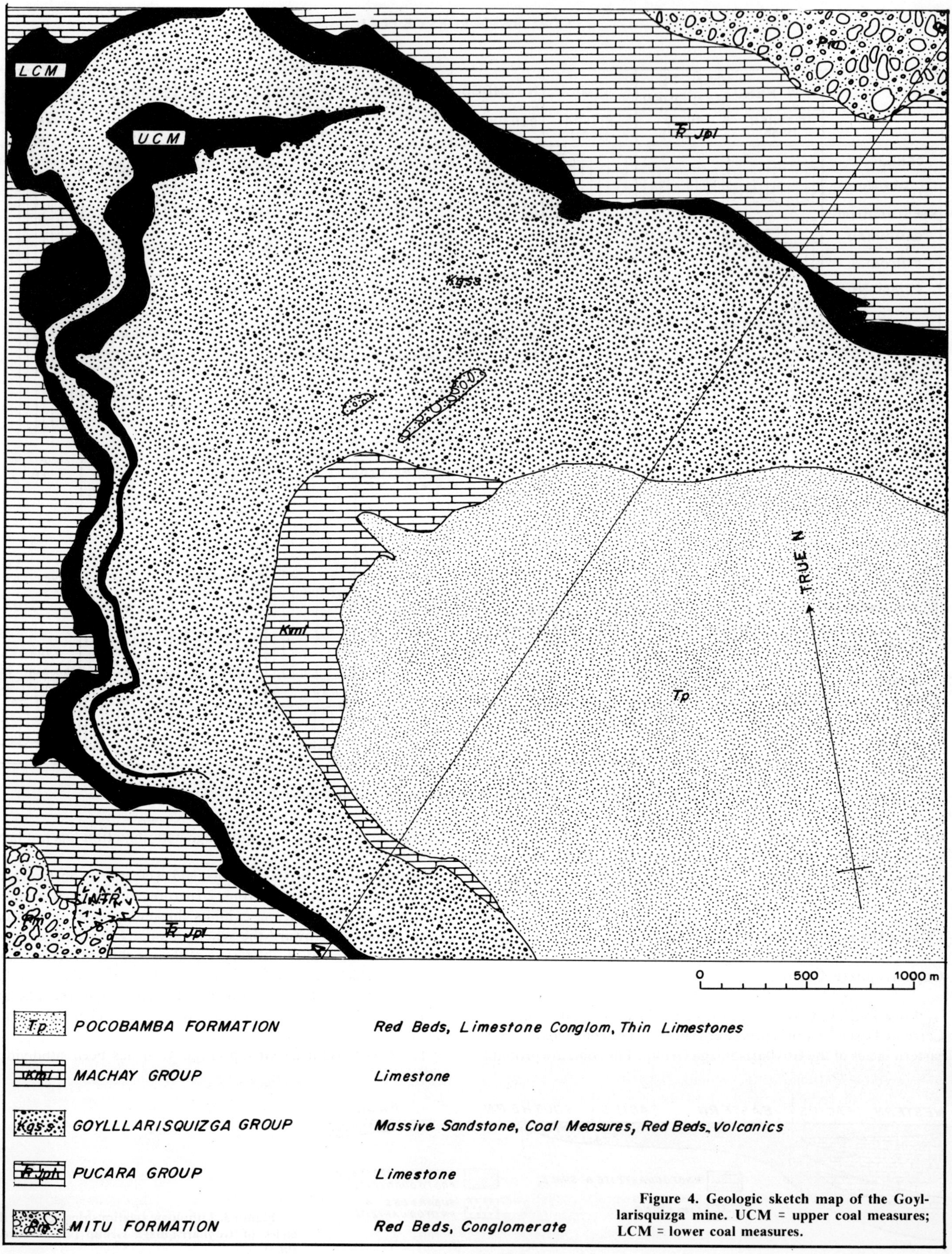

Figure 4. Geologic sketch map of the Goyllarisquizga mine. UCM = upper coal measures; LCM = lower coal measures.

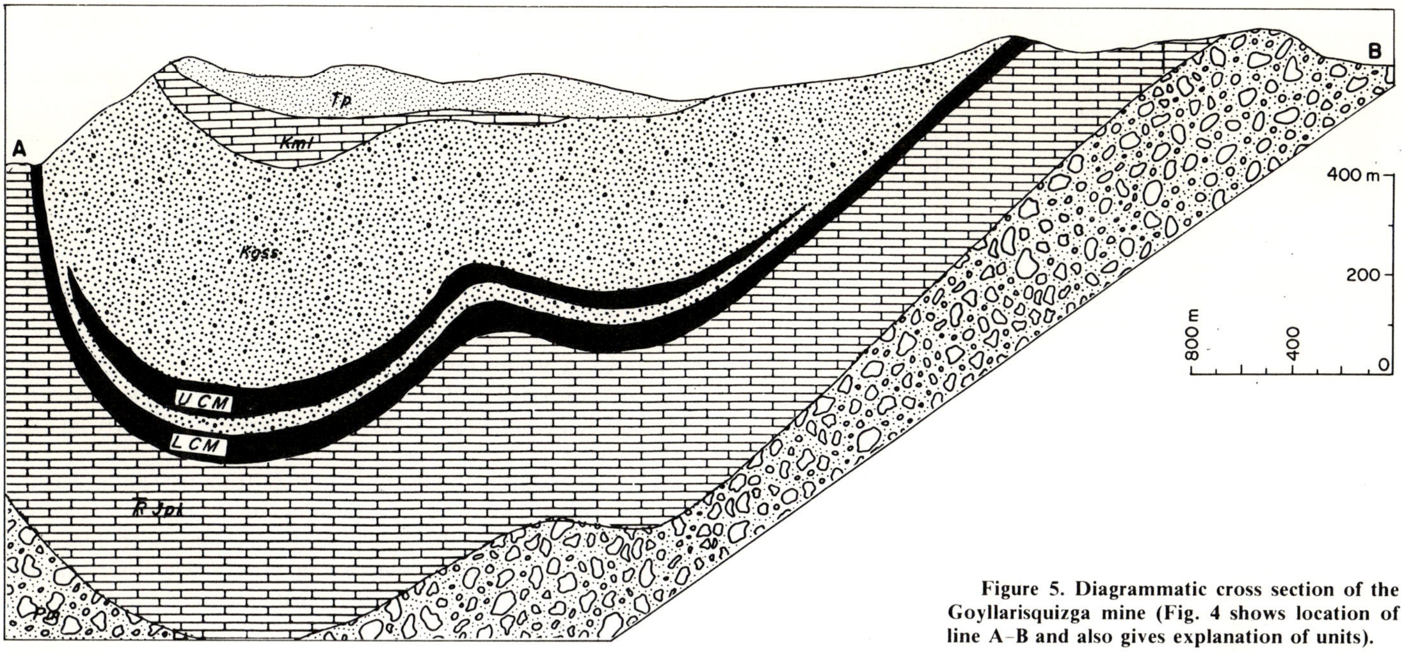

Figure 5. Diagrammatic cross section of the Goyllarisquizga mine (Fig. 4 shows location of line A–B and also gives explanation of units).

informally at this locality, as follows, from top to bottom: calcareous sandstone, Chontas Sandstone, Bolognesi beds, upper coal measures, Murucata Sandstone, and lower coal measures.

The two upper sandstone members combined are 400 to 500 m thick. The Bolognesi beds are 100 to 150 m thick. They consist of a sequence of reddish shale and sandstone with interbedded amygdaloidal andesitic to basaltic flows.

The upper coal measures consist of sandy fire clay, mudstone, and bituminous shale. Locally, five coal beds of mineable thickness have been found. These wedge out on the flanks of the syncline. Volcanic tuff beds occur at the base of this unit.

The Murucata Sandstone is about 90 m thick along the synclinal axis, but less than 30 m thick on the flanks of the syncline.

The lower coal measures are extremely variable in composition, which indicates an unstable depositional environment. This sequence includes only one fairly persistent coal bed. A conglomerate bed is at the base. No volcanic rocks have been seen.

Mining Operations. The Goyllarisquizga mine was owned and operated by the Cerro de Pasco Corporation, and the coal was used entirely to produce coke for the metallurgical operations of the corporation. The coal in the Goyllarisquizga field was deposited under unstable conditions in an intermontane basin and, consequently, has a large ash content. The coal produced in 1967 averaged 54 percent ash. Argillaceous bands are characteristic in the coal bed. The coal ranges in rank from subbituminous to bituminous. Certain beds or layers yield a satisfactory metallurgical coke that has a high ash content and a low carbon content.

Like most Peruvian coal, the coal in the Goyllarisquizga field has been deformed tectonically. As a consequence, the coal breaks into rhombic flakes as it is mined. This phenomenon is observed even in areas of hard coal, which has to be mined with explosives, because the coal breaks along pre-existing shear planes.

Most of the recent production came from a southeasterly extension of the coal beds; this new extension is called the El Dorado section. This section was discovered during an exploration program in 1963, 1964, and 1965.

In some parts of the El Dorado section, two beds are known to exist. No correlation has been made with beds in the longer producing parts of the field. The two beds are separated by a band of clay, brownish mudstone, and silty shale. Almost all development in the El Dorado section has been in the upper bed, which is overlain by dark carbonaceous shale locally called "Caja Negra."

A characteristic in some zones of the El Dorado section is the presence of large lens-shaped concretions called "muqui" by the local miners. These concretions are as much as 6 m long and 0.5 to 1.0 m thick. The muqui constituted the main obstacle to mechanization in the mine because they are very hard, and the larger ones required blasting. Microscopic examination shows that these concretions consist of a fine-grained mixture of calcite and iron pyrite.

Washing and Coking. The washing and coking process is of considerable interest in view of the low quality of the coal mined.

The coal washery was a simple gravity-concentration plant. The raw coal was broken first in a hammer mill working in closed circuit with a 6 by 10 ft vibratory screen (fitted with a 0.5-in. opening cloth). The coal pieces of less than 0.5-in. size were then fed successively into two three-compartment rougher jigs, two three-compartment scavenger jigs, and one four-compartment rescavenger jig. The float product from the final or rescavenger jig was classified as refuse. The first-class coal was stored in wooden bins for dewatering before it was shipped to La Oroya for subsequent coking. An average of 13,000 tons per month of raw coal was treated in the washery. The average plant feed assays of the raw coal are shown on Table 1, together with the analysis of the washed, first-class coal. Yield or weight recovery of the washed product (shown in Table 1) is 38.6 percent.

The washed coal was transported by rail to La Oroya. The coal was mixed at La Oroya with the fines that resulted from the production and handling of the coke. The proportion of fines to be added to the coal was about 20 percent. The coking cycle was normally 15 hr, and the analyses of the product are shown on Table 1. The coke produced in La Oroya had a larger ash content than

TABLE 1. QUALITY OF COAL, GOYLLARISQUIZGA MINE, PERU

Components	Raw (%)	Washed (%)	Coked (%)
Volatile matter	23.4	30.4	4.0
Fixed carbon	22.6	43.1	54.1
Ash	54.0	26.5	41.9
Sulfur	12.1	3.3	2.9

TABLE 2. QUALITY OF COAL, JATUNHUASI MINE, PERU

Components	Raw (%)	Washed (%)	Coked (%)
Volatile matter	27.6	34.3	1.4
Fixed carbon	30.1	39.0	59.8
Ash	42.3	26.7	38.8
Sulfur	5.1	2.7	2.2

cokes produced elsewhere in Peru and possibly elsewhere in the world. The total production of about 1,700 tons per month was used in the lead furnaces.

Jatunhuasi Coalfield

Location. The Jatunhuasi coalfield is about 70 km south of La Oroya, close to the crest of the Andean Cordillera but east of the continental divide (Fig. 1). The field is in the southern facies of the Goyllarisquizga Group, which at this locality is about 400 m thick.

Geology. The main outcrop of the coal-bearing beds in the field lies along the eastern limb of a broad syncline with beds extending across a strike distance of 30 km. The strike undulates gently, from N30°W to N85°W, with a regular southwest dip averaging about 30° (Fig. 6). The structure is regular with beds unbroken for miles except for a few cross faults with little displacement.

Six coal beds crop out in most parts of the field, but only two have the quality and thickness to be considered workable. The coal probably was deposited in a deltaic environment.

Mining Operations. The Jatunhuasi field was previously developed on a small scale by shallow workings, and in recent years it has been explored at depth by the Cerro de Pasco Corporation. At the present time, the Jatunhuasi field can be considered to have a resource potential similar to that of the Goyllarisquizga field, that is, approximately 8 million metric tons of coal. However, very little of this estimated total has been proved.

Because the field has not been developed, no average figures for the quality of the coal are known. During 1955, a sample of about 5,000 metric tons was mined from a section of the field. The raw coal was shipped to the washery, and the clean coal was returned to La Oroya for coking. The results are shown on Table 2.

The coke was hard and blocky, similar to that made from Goyllarisquizga coal. No significant differences between Jatunhuasi and Goyllarisquizga cokes appeared during the full-scale plant run in the lead furnace.

Oyón-Checras Coalfield

Location. The Oyón-Checras coalfield is in the western facies of the Goyllarisquizga Group. It extends from 10 km north of Oyón to about 10 km south of Parquín (Checras), parallel to and west of the continental divide (Fig. 1).

Geology. The western facies of the Goyllarisquizga Group has been subdivided into the following formations (in ascending order): Oyón, Chimú, Santa, Carhuaz, and Farrat Formations. The coal in this field is in the Oyón Formation, which consists of about 100 m of thinly bedded subgraywacke, shale, and lenticular coal beds that were probably deposited in a swamp. The age is probably early Valanginian or Berriasian (earliest Cretaceous), but a Tithonian age (latest Jurassic) cannot be completely excluded. The base of the formation is not exposed (Fig. 2).

Because of intense folding, most of the coal beds dip 40° or more, except where the anticlinal summits are exposed. In the anticlinal crests, the coal generally is squeezed into thick pockets of crushed coal, commonly mixed with material from overlying and underlying beds (Fig. 7).

The Oyón-Checras coals range in rank from true anthracite to medium volatile bituminous coal. Sulfur content is low everywhere in the field. The coal in general is friable and, with repeated handling, breaks into fines. Ash content is not finely disseminated.

Mining Operations. The area most intensely explored is a bituminous deposit at Conocpata, near Oyón. This exploration was done by the Banco Minero del Peru, whose primary object was to obtain coke for domestic use in Lima and other cities, as well as coal for export. This work was done between 1943 and 1946, but results were unsatisfactory. Average analysis of Conocpata coals in channel samples is 16.3 percent volatile matter, 66.2 percent fixed carbon, 16.0 percent ash, and 1.2 percent sulfur.

Outside the Conocpata area, the small amount of information available on the physical character of the coals indicates that they are mainly anthracite.

The total possible tonnage that could exist in the Conocpata area is about 8 million tons, but an economic method for mining this steeply dipping and friable coal is still an unsolved problem.

Northern Anthracite Field

The Northern Anthracite field extends from the area of the upper Chicama River in the north about 250 km southward to the Santa River valley (Fig. 1). This field has the largest potential for anthracite in Peru, but present development is on a small scale because of the lack of a market. The coalfield is in the western facies of the Goyllarisquizga Group, but unlike the Oyón-Checras field, the coal is mainly in the Chimú Formation with some beds in the Carhuaz and Farrat Formations.

Upper Chicama River Area. The northern part of the Northern Anthracite field, which comprises the upper Chicama River area, has been explored recently by the Peruvian Government through its Servicio de Geología y Minería. The coal beds are found within the Chimú Formation, which is similar in lithologic character, sedimentary structure, and fossil content to the eastern facies of the Goyllarisquizga Group and therefore is believed to have been deposited in the same deltaic and flood-plain environments.

Estimated reserves in the mined and explored areas are small compared with the total resource potential of the area. Potential resources are considered to be about 200 million tons, according to the studies by the Servicio de Geología y Minería (J. Escudero,

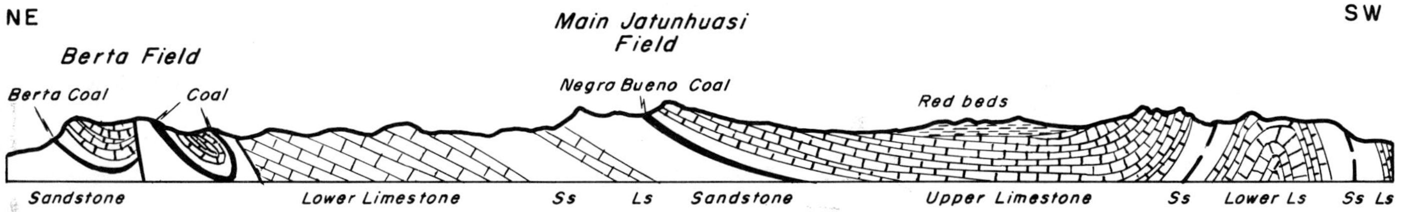

Figure 6. Diagrammatic cross section through Jatunhuasi district, viewed toward the southeast. No scale. Length of section 10 to 12 km (after McLaughlin, 1922). Ls = limestone; Ss = sandstone.

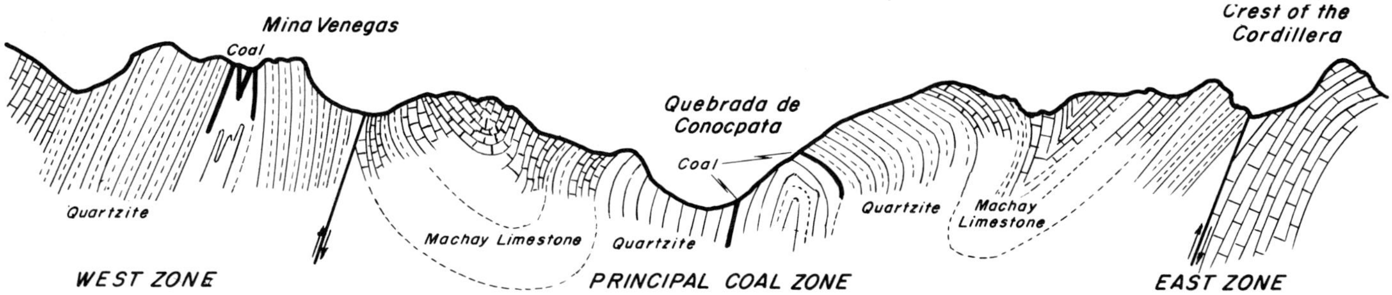

Figure 7. Diagrammatic cross section across Quebrada de Conocpata, viewed toward the northeast. Oyón coalfield (after McLaughlin, 1921).

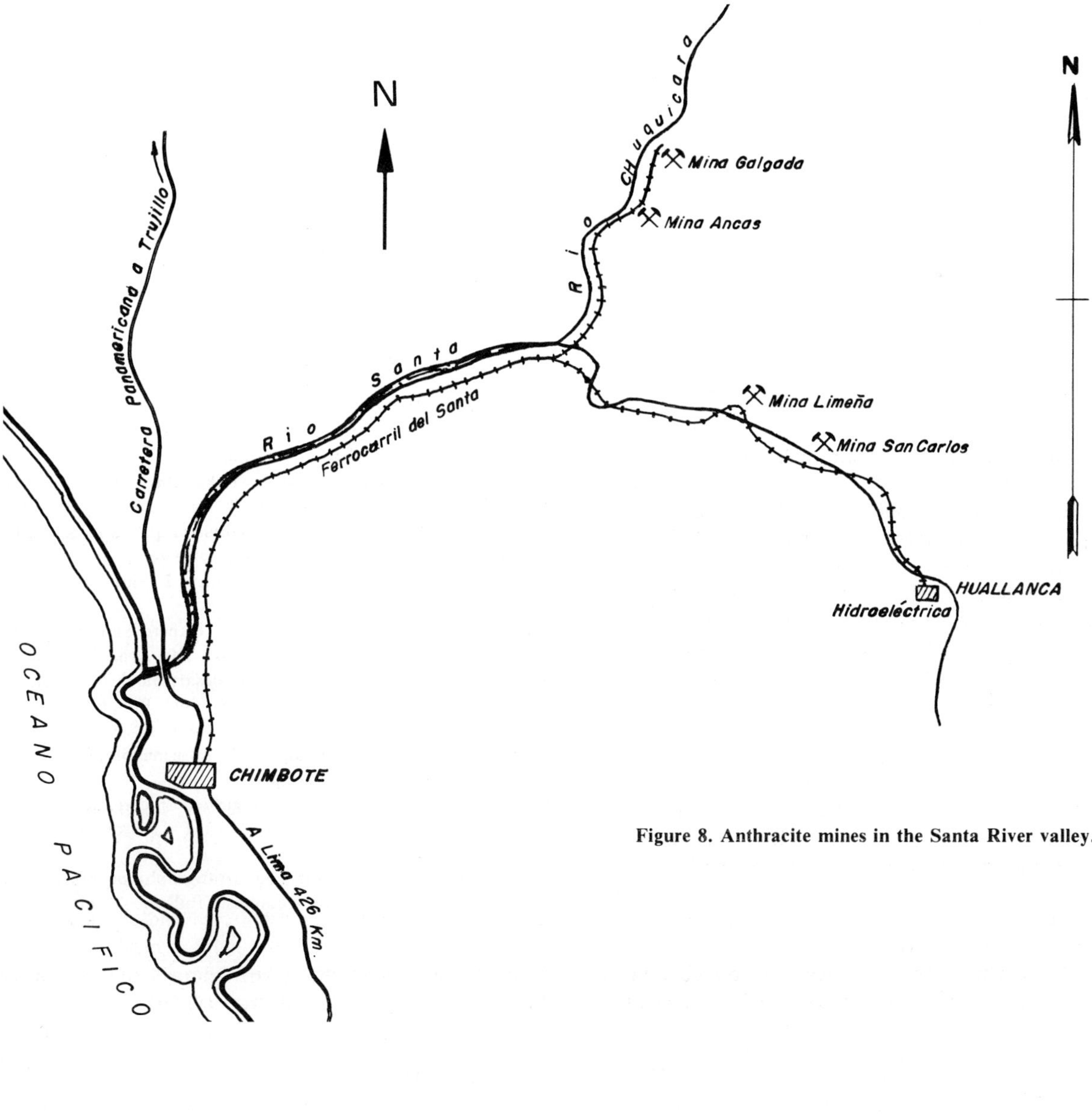

Figure 8. Anthracite mines in the Santa River valley.

1968, oral commun.). It is of some interest to compare this quantity with the potential estimated by Richard Stappenbeck, a German geologist who worked in the area before 1929. For the same region he estimated a potential of 170 to 180 million tons of recoverable coal.

The information on the quality of the coal is insufficient to give an average composition for the field, but the fixed carbon content ranges from about 80 to 90 percent and the volatile matter from about 2 to 10 percent.

Santa River Valley Area. The coal deposits in the Santa River valley have been studied repeatedly for the Corporación Peruana del Santa (a government institution) in efforts to justify increased production. This area has the advantage of being easily accessible from the Chimbote steel works and port (Fig. 1).

The coal beds are strongly folded and faulted, as are most beds west of the continental divide. The summits of the anticlines have been eroded and the coal is found on the flanks dipping at angles of 50° to 90°. The coal is deeply weathered and is not commonly recognizable as coal in the surface exposures, because it has been oxidized to a residual ash. The weathering extends downward for 30 to 50 m. The average composition of the coal beds is as follows: fixed carbon, 75 percent; volatile matter, 4 percent; ash, 16 percent; and sulfur, 1 percent.

Several mines have been active in this area. The most important are the Galgada, Ancas, Limeña, and San Carlos mines (Fig. 8). The maximum resource potential of the area in which these mines are located amounts to 50 million tons, but because of little exploration and development, the actual proved resources are small.

ECONOMIC CONSIDERATIONS

The only major consumer of bituminous coal in Peru is the Empresa Minera del Centro del Peru (Centromin), which uses it for making metallurgical coke. The reserves and total resource potential of this type of coal, although relatively small, would be sufficient for many years at the present rate of consumption, provided that mining could be economically done.

Potential resources of anthracite are great, but the problem of a suitable market is a serious one. Studies have been made of the feasibility of using this coal in the steel works at Chimbote, as fuel in thermoelectric plants, and as a blend with imported bituminous coal in the production of coke. None of these uses was practicable because of strong economic competition from fuel oil and hydroelectric power and because of the relatively low quality of the coal, especially its low content of volatile matter.

CONCLUSIONS

Peru's potential resources of bituminous coal are approximately 20 million metric tons in deposits in the central part of the country, but exploitation is strongly dependent on the development of an economic mining method.

The potential tonnage of anthracite is about 250 million tons, located mainly in northern Peru.

Although the quality of the Peruvian bituminous coal is rather poor, it yields an acceptable metallurgical coke that can be used in the metallurgical operations in central Peru (for example, Centromin, La Oroya).

The quality of the anthracite is such that no important use for it has been developed so far.

ACKNOWLEDGMENTS

Acknowledgment is due to the management of the Cerro de Pasco Corporation for making possible this review and for authorization to use information from its private files, especially in regard to the Goyllarisquizga, Jatunhuasi, and Oyón coalfields. Ing. J. Escudero, of the Servicio de Geología y Minería, contributed information on the recent work in the upper Chicama River valley area. Ing. Alfredo García Stens of the Banco Minero del Peru helped with discussions and information from the files of the institution. A.A.G. Bowerman offered helpful suggestions in the preparation of the final text.

SELECTED REFERENCES

Benavides-Cáceres, V., 1956, Cretaceous system in northern Perú: Am. Museum Nat. History Bull., v. 108, art. 4, p. 252–494.

Bennet, Evan, 1951, Report of coal investigations: La Oroya, Perú, Cerro de Pasco Corp., unpub. private rept.

——1952, Coal report: La Oroya, Perú, Cerro de Pasco Corp., unpub. private rept.

Fischer, A. G., 1956, Desarrollo geológico del noroeste peruano durante el Mesozóico: Soc. Geol. Perú Bol., t. 30, p. 177–190.

Gerbracht, F. P., 1958, Coal and the Jatunhuasi program: La Oroya, Perú, Cerro de Pasco Corp., unpub. private rept.

Harris, L., and Quevedo, F., 1967, Aprovechamiento de los finos en la producción de coke metalúrgico: Lima, Perú, 10th Convención Nac. Ingenieros Minas.

Jenks, W. F., 1948, Geology of the Arequipa quadrangle: Inst. Geol. Perú Bol., t. 9, 204 p.

Kummel, B., 1948, Geological reconnaissance of the Contamana region, Perú: Geol. Soc. America Bull., v. 59, p. 1217–1265.

McLaughlin, D. H., 1921, Report on the Oyón coal field: La Oroya, Perú, Cerro de Pasco Corp., unpub. private rept.

——1922, Report on the Jatunhuasi coal field: La Oroya, Perú, Cerro de Pasco Corp., unpub. private rept.

Newell, N. D., 1949, Geology of the Lake Titicaca region, Peru and Bolivia: Geol. Soc. America Mem. 36, 111 p.

Page, J. S., 1962, Coal inventory, Goyllarisquizga mine: La Oroya, Perú, Cerro de Pasco Corp., unpub. private rept.

Rivera, R., 1951, La fauna de los estratos Puente Inga, Lima: Soc. Geol. Perú Bol., t. 22, 53 p.

Ruegg, W., 1957, Geologie zwischen Cañete–San Juan, 13° 60′–15° 24′, Südperu: Geol. Rundschau, Bd. 45, Hft. 3, p. 775–858.

Rumscheidt, W., and others, 1956, El desarrollo de la industria carbonífera y siderúrgica de Chimbote, Perú: Lima, Banco Minero del Perú, unpub. private rept.

Sofremines (consulting firm), 1958, Informe preliminar: Lima, Banco Minero del Perú, unpub. private rept.

Stappenbeck, Richard, 1929, Geologie des Chicamatales in Nord-Perú und seiner Antrazit Lagerstätten: Geol. Paläontol. Abh. v. 16, no. 4, p. 1–51.

Wilson, John J., 1963, Cretaceous stratigraphy of central Andes of Perú: Am. Assoc. Petroleum Geologists Bull., v. 47, no. 1, 34 p.

Manuscript Received by the Society March 10, 1975
Manuscript Accepted March 17, 1975

Printed in U.S.A.

Geological Society of America
Special Paper 179

Coal basins of Ecuador

J. E. O'Rourke
113 Sylvania Drive, Pittsburgh, Pennsylvania 15236

ABSTRACT

The coal deposits of southern Ecuador are in remnants of Tertiary basins now perched in topographic basins at elevations of 2,000 to 3,000 m. Recent mapping suggests that undisturbed coal beds may underlie broad areas, but they will remain untouched because the cover is thick. The known seams, exposed in narrow faulted belts, are steep, disrupted, and shattered. They are as thick as 5 m and contain mostly subbituminous coal and lignite, but have objectional amounts of shale, bentonite, and pyrite. Reserves in the exposed seams of the Malacatus and Loja basins are said to be roughly 1 million metric tons each; in the Cañar-Azuay basin, the reserves are said to be roughly 20 million metric tons. Little of this can be considered economically mineable, however, because an attempt at semimechanized underground mining was unsuccessful. Subterranean gasification may be required.

INTRODUCTION

The coal resources of Ecuador have been investigated in the cooperative programs of several international agencies from 1958 to 1968, but very little information has been published. Results of the most comprehensive mapping project, sponsored by the United Nations Development Programme and the Servicio Nacional de Geología y Minería, have been circulated among their offices as a mimeographed report (United Nations and Republic of Ecuador, 1969). The present paper is based largely on this field work.

MINOR DEPOSITS

Small lignite bodies have been noted in Miocene rocks of the Amazon basin and in the Oligocene Series of the Pacific coast, but the only deposits of economic interest are found in the intramontane basins of the Andes, which in Ecuador are divided into the Cordillera Occidental and Cordillera Oriental (or Real; Fig. 1).

An occurrence at San Antonio de Pichincha, 30 km north of Quito, is the only known example of lignite younger than Tertiary. The seam, 2 to 4 m thick, interlayered with silt, is exposed in the Río Monjas at the municipal swimming pool and on the west bank about 1,800 m upstream. Coarse, woody debris seems to have been dumped into an early Pleistocene(?) lake measuring roughly 2 by 10 km. A cover of loose pyroclastic deposits 15 to 20 m thick precludes mining.

Lenses of coal have been reported from Columbe, Chimborazo Province, and from Nabón, about halfway between Cuenca and Loja, but are said to be too small for mining (Putzer and Schneider-Scherbina, 1958).

MALACATUS BASIN

The southernmost coal basin surrounds the village of Malacatus (or Malacatos; lat 4° 08′S, long 79° 13′W), about 35 km by unpaved road south of Loja (Fig. 1). A sketch map of the area at a scale of 1:43,000 has been made by Alvarado (1967). It shows a belt of Tertiary sedimentary rocks about 5 km wide stretching northwestward for about 25 km between hills of metamorphic and volcanic rocks. Beds of conglomerate, siliceous shale, siltstone, and sandstone with traces of hydrocarbons dip moderately eastward, except near faults, where they are steep. Flat-lying Quaternary gravel and pyroclastic rocks obscure these strata in many places. Miocene ostracods have been found at the northern end of the basin by geologists of the Institut Français du Pétrole.

Coal seams exposed in the upper part of the section have been mapped and sampled by German and French missions (Putzer and Schneider-Scherbina, 1958; Spindler and others, 1959). The seams reach 2 m in thickness, but are lenticular and disrupted by faults (Fig. 2). Mining has been limited to pick-and-shovel work. Reserve estimates range from 712,500 to 3,290,000 metric tons (hereafter, t). Seven samples average 4,500 cal/g, 15 percent ash, 8 percent sulfur, and 2 percent free moisture.

LOJA BASIN

The small city of Loja (or Loxa; lat 4° 00′S, long 79°12′W) is in the center of a Tertiary basin about 10 km wide, 20 km long, and 2,200 m high. The enclosing mountains are carved from metamorphic rocks of undetermined age. The general geology has been mapped at a scale of 1:43,000 (Alvarado, 1967). The sedimentary rocks of the basin dip slightly to moderately eastward, except in a central strip about 1,500 m wide where they are strongly faulted and

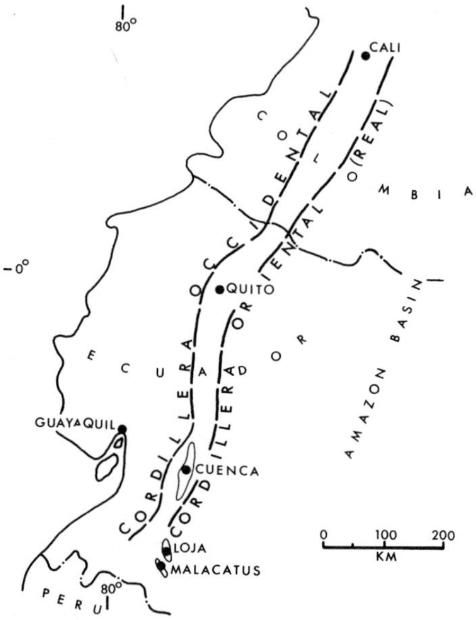

Figure 1. Index map showing location of Tertiary basins in Ecuador.

dip steeply. Conglomerate, sandstone, and shale predominate. Minor rock types include siliceous shale, dolomitic marl, and a distinctive, white, slightly consolidated sediment that averages 84 percent silica (as very fine alpha quartz), has a specific gravity of 0.7, and contains fossil fish. Similar material, also with fish, occurs in the Oligocene Dos Bocas Formation at Sucre, Province of Guayas, on the Pacific coast (also see Canfield, 1966, p. 57).

The coal seams exposed in the central belt are 1 to 2 m thick and have been mined on a small scale from numerous adits, now caved in. Reserve estimates range from 593,000 t (Spindler and others, 1959) to 3,780,000 t (Putzer and Schneider-Scherbina, 1958). Three samples from the best seam average 5,200 cal/g, 9 percent ash, 8 percent sulfur, and 10 percent moisture, but 11 samples from the others have a mean of 3,900 cal/g, 30 percent ash, 7 percent sulfur, and 7 percent moisture.

CAÑAR-AZUAY BASIN

The largest of the Andean Tertiary basins, measuring roughly 20 by 100 km, extends across the provinces of Azuay and Cañar and centers around the city of Cuenca (Fig. 1; lat 2° 52′S, long 78° 59′W). It is the only coal basin served by a railroad, and the only one in which coal has been mined for industry. I mapped the area on a scale of 1:250,000 in 1967. The coal district was mapped on a scale of 1:50,000 and the locality of active mining on a scale of 1:10,000 by A. Schneider-Scherbina and C. F. Mosquera in 1965 (unpublished).

The first petroleum geologists to examine the area were handicapped by a lack of base maps and air photos. Each in turn applied the formation names Azogués, Biblián, and Cuenca to different units. Also, they were not aware that the basin is cut through the center by a major fault, so that the folded rocks on the eastern side represent only the lower half of the Tertiary column, whereas the flat beds on the western side are the upper half (Figs. 3, 4).

Stratigraphy

The pre-Tertiary rocks in the Cordillera Occidental resemble the volcanic rocks of the Lower Cretaceous Piñon Formation of the coast (Sauer, 1971, p. 52). Those underlying the basin in the Cordillera Oriental include black, pyritic, bituminous argillite; black limestone; black chert or lydite; gray, laminated, fine-grained sandstone; pyroclastic rocks; and mafic sills. These rocks had been overlooked or assigned to the "pre-Mesozoic basement complex." However, they are similar to those dated as Maastrichtian (latest Cretaceous) west of Quito, and to much of the Napo Group in the Amazon basin. Also, I found casts of *Inoceramus* at two places east and southeast of Azogués (Table 1; Fig. 5).

The overlying unit is not penetrated by the mafic intrusive rocks and contains pebbles of the rock types just mentioned. It consists of red or purple, silty, ferruginous (as much as 21 percent Fe_2O_3) clay, with lentils of gray, cross-laminated, granule conglomerate. It has yielded no fossils, but resembles the red beds of the Río Saloya, southwest of Quito, and is analogous to the Tena Formation of the Amazon oil fields, generally placed in the interval between uppermost Cretaceous and middle Eocene.

The next younger formation overlaps this basal unit and in several places overlies directly the Mesozoic rocks. The lower fourth of the unit is mostly sandstone and shale with thin layers of pebble conglomerate. Poorly preserved gastropods have been identified by D. Mongin (1968, written commun.) as *Turritella* aff. *T. chira* Olsson, said to occur in the upper Eocene brackish-water sedimentary rocks of Colombia and Peru. The upper 200 m of this formation is black pyritic shale, which is bleached, coated with limonite, alum, and jarosite, and veined by selenite. A few lenses of black, chalcedonic limestone, as much as 0.5 m thick and 2 m long, are enclosed.

The next younger unit overlaps the two lower formations at several places. It is a massive, tuffaceous sandstone, 50 to 500 m thick, with thin beds of siltstone and shale. Pelecypod casts found by E. Núñez del Arco near its base were identified tentatively by Mongin (1968, written commun.) as *Nucula* cf. *N. andersoni* Clark (1946).

A black pyritic "paper shale," with beds of pure calcium bentonite as much as 10 m thick, is exposed between Guapán and Charasol. Plant fossils are well preserved. The large size of both the

Figure 2. Malacatus basin; coal seam is 2 m thick (photo by Schneider).

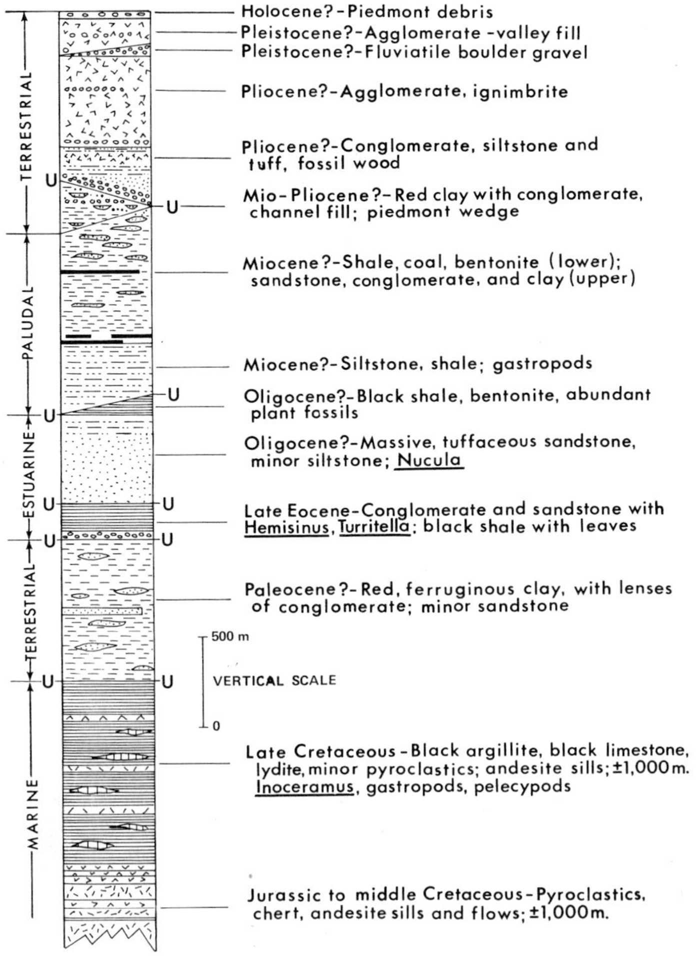

Figure 3. Stratigraphic column, Cañar-Azuay basin. U, unconformity.

TABLE 1. NEW FOSSIL LOCALITIES, CAÑAR-AZUAY BASIN

Age	Location	Type
Cretaceous	San Cristobal road, 2 km northeast of Descanso	*Inoceramus*
	San Marcos road, 1 km east of Azogués	*Inoceramus*
	Jadán road, Quebrada Guava-huaico	Foraminifera
	Tabacay road, 4 km north of Azogués	Gastropods
	300 m west and 1,000 m south of Cumbe	Pelecypods
Eocene(?)	Railroad bridge, 2.5 km northwest of Azogués	*Turritella*
	Chuchingay, 4 km northwest of Azogués	Flora
Unknown	San Miguel (500 m northwest), near road	*Nucula*
	Guapán, east of cement plant	Flora
	Charasol, 500 m southeast of guard post	Flora
	Cushumaute, 2 km north of San Nicolás	Gastropods
Miocene(?)	Cochahuaico, 1.5 km west of Biblián (adit)	Pelecypods
	Cushumaute mine, 1.5 km north of San Nicolás	Flora
	Curiquingue, 2.5 km west of Cojitambo	Flora
25,000 yr B.P.	Gualaquiza road, 40 km south of Gualaceo	Charcoal
Holocene	Guapán travertine quarry	*Homo sapiens*

leaves and the leguminous seed pods suggests a tropical environment.

A formation with alternate layers of siltstone and mudstone, arranged in large-scale cross-bedding, is the next overlying unit. At least 500 m is exposed, but the lower contact is obscured by a thrust fault.

Two zones of coal seams (discussed below) are contained in an interval of clastic sedimentary rocks, mostly shale, siltstone, and bentonite. The complete thickness cannot be measured because of the central fault, but is not less than 500 m. Stems, branches, and leaves can be discerned in the lignitic facies. A coquina of pelecypod shells overlies one coal seam at Cochahuaico. "Crocodile" teeth and other vertebrate remains have been reported by the villagers.

Coarse tuffaceous sandstone and conglomerate layers are more common above the coal zones, and a unit of red beds with lenticular conglomerates overlies them on the west. These coarse layers presumably represent the onset of the Andean orogeny, usually dated as Miocene-Pliocene. Intensely altered volcanic rocks rim the basin and cap some of the interior mesas, but there are no volcanoes here as in northern Ecuador.

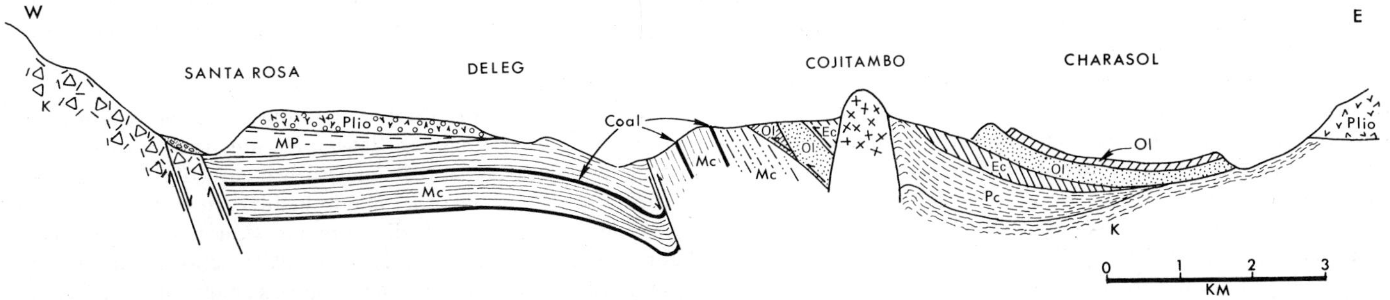

Figure 4. Diagrammatic cross section of Cañar-Azuay basin. K, Cretaceous; Pc, Paleocene(?); Ec, Eocene; Ol, Oligocene(?); Mc, Miocene(?); MP, Miocene-Pliocene; Plio, Pliocene.

Figure 5. *Inoceramus* cast, San Cristobal road.

Coal Deposits

The coal deposits have traditionally been grouped by the miners into the "Washington zone," 50 m thick, with three to four seams of workable size; and the "Cañari zone," about 500 m higher, which has one main bed 2 to 5 m thick, plus several thinner layers. All are lenticular.

The thin seams are mostly lignite, whereas the thick ones are mostly coal, apparently vitrain and fusain. The product of the San Luis mine was described as subbituminous in rank (Pierce Management Corp., 1966). The weathered material, which extends to depths as great as 10 m, contains selenite plates, limonite stains, and iron sulfate coatings.

The seams are best developed in the vicinity of Biblián, about 20 km northeast of Cuenca. On the north they are too thin, and on the south they have too much ash. For example, 15 samples from the Washington zone near Biblián average 4,400 cal/g, 21 percent ash, 6 percent sulfur, and 6 percent moisture, whereas 11 samples from this zone farther south have a mean of 2,100 cal/g, 45 percent ash, 6 percent sulfur, and 11 percent moisture. Five samples of the Cañari seam in the southern area, near the exposure shown in Figure 6, average 2,400 cal/g, 41 percent ash, 6 percent sulfur, and 11 percent moisture. The fact should be noted, however, that much of the ash is present as thin layers of bentonite and shale that might be removed by washing.

Reserves

The appraisal of reserves is hard because of lenticularity, faults, and lack of outcrops. Several attempts have been made (Mosquera, 1950; Bennett, 1952; Putzer and Schneider-Scherbina, 1958). These calculations assume an average thickness for the seams and multiply this figure by the strike length and mineable depth. They range from 15,750,000 to 22,550,000 t.

The possibility that the coal seams may be buried in the western half of the basin, an area of about 150 km^2, was tested during the U.N.-SNGM project. The presence of an outcrop of good coal in a faulted wedge near Chiquintad, at the foot of the Cordillera Occidental, and of another in the core of an anticline on Río Sidcay, just west of the central fault, suggests that the coal measures may persist beneath the flat-lying postorogenic conglomerates and volcanic rocks. Funds were sufficient to drill two rotary drill holes. The one at Deleg went to a depth of 363 m, and the other at Burgay went to a depth of 325 m. Neither reached the Canari seam.

Semimechanized underground mining was attempted in the mid-1960s at the San Luis mine in the Biblián district. The cost of timbering the shattered wall rock and of driving numerous crosscuts to recover faulted seams was excessive. The coal had been used to roast travertine at the nearby cement plant run by the same company, Empresas Industrias Guapán, S.A. When the discovery of oil in the Amazon basin made that fuel cheap, the mine was abandoned.

CONCLUSIONS

The Tertiary basins of southern Ecuador have now been mapped at medium scale, and at least the main features of their geologic setting are understood. No marker horizon, which would show a connection among the three basins, has yet been found, and the fossil age determinations are not conclusive, but the evidence indicates that the coal is Miocene or perhaps a little older.

The sum of the maximum thickness of the Tertiary sedimentary formations in the Cañar-Azuay basin is about 3,000 m, much greater than previously thought. Several hundred metres of black shale and other fine-grained sedimentary rocks below the coal include remains of what appear to be brackish-water mollusks and tropical plants. This new evidence supports the interpretation of an estuarine environment like that inferred for the Colombian coal basins or approximately like that of Guayaquil Gulf today (A. A. Olsson in a letter quoted in Liddle and Palmer, 1941, p. 20-21). Thus, the southern basins seem to be poorly exposed and incompletely preserved parts of extensive older Tertiary deposits, not just freshwater lakes filled during the main Andean orogeny.

The structure of the three basins also supports this interpretation. Broad areas of flat-lying conglomerate or gravel, interlayered with pyroclastic rocks, are interrupted by narrow belts of steep, fine-grained sedimentary deposits containing the coal. The older Tertiary rocks are exposed only where they have been deformed, elevated, and stripped of the continental debris.

Any estimate of the total coal resources of the region must consider that the tonnage in flat-lying seams concealed beneath Pliocene and younger rocks in undisturbed parts of the basins may be considerable. On the other hand, the first attempt to trace such deposits by deep drilling was discouraging. Hence, the only reserves that can be counted are the relatively small fractions in the steeply dipping seams. Ironically, this coal was so strongly shattered by the faulting that brought it within reach that it too cannot be mined.

The only remaining possibility is that the coal might be

Figure 6. Cañari coal seam at Ayancay, east of Cuenca (photo by Schneider).

exploited by some unorthodox method. For example, the properties that have proved detrimental to ordinary mining (steep dips, shattering, high ash, high volatility) are advantages in underground gasification.

Finally, new evidence for the possible presence of hydrocarbons in the Cañar-Azuay basin should perhaps be mentioned. Fault-controlled travertine deposits, mapped by myself and discussed in an unpublished report, indicate that the basin is underlain by Cretaceous limestone. Widely scattered asphalt veins, formerly thought to be distillations from the coal, are now known to cut the oldest Tertiary sedimentary rocks and are more likely to have come from the Upper Cretaceous rocks, which resemble the source rocks in the Amazon basin and contain sufficient organic matter to generate hydrocarbons.

REFERENCES CITED

Alvarado, R., 1967, Estudios geológicos de una cuenca sedimentaria de los carbones del Austro [thesis]: Guayaquil, Ecuador, Escuela Politécnica del Litoral, 61 p.

Bennett, E., 1952, Proyecto de explotación del lignito de Biblián, Ecuador: U.N. Tech. Assistance Program, 33 p.

Canfield, R., 1966, Geologic report of the coast of Ecuador: Quito, Asesoría Técnica Petróleos Ministerio Industrias y Comercio, 137 p.

Liddle, R. A., and Palmer, K.V.W., 1941, The geology and paleontology of the Cuenca-Azogués-Biblián region, Provinces of Cañar and Azuay, Ecuador: Am. Paleontology Bull., v. 26, no. 100, p. 360–421.

Mosquera, C. F., 1950, Los yacimientos carboníferos de la Provincia de Cañar: Quito, Bol. Inf. Cien. Nac., v. 3, no. 35, p. 321–355; no. 38, p. 631–635.

Pierce Management Corporation, 1966, The San Luis Mine, Biblián: Scranton, Pa.,Pierce Management Corp. (unpub. consultant's rept.), 30 p.

Putzer, H., and Schneider-Scherbina, A., 1958, Lagerstätten-Untersuchungen in Ecuador: Quito, Misión Alemana Coop. Tecnica Minera, 30 p.

Sauer, W., 1971, Geologie von Ekuador: Berlin, Borntraeger, 316 p. (includes a summary of coal resources by H. Putzer, p. 212–221).

Spindler, J. P., Mangez, G., Mosquera, C. F., and Herrera, J. I., 1959, Los carbones de Malacatos y Loja: Quito, Misión Geol.-Minera Franco-Ecuatoriana, Inf. 3, 29 p., 2 maps.

United Nations and Republic of Ecuador, 1969, Survey of metallic and non-metallic minerals: New York and Quito, Tech, Rept. No. 1, Coal Investigations, U.N. Devel. Programme and Servicio Nac. Geología y Minería, 59 p., 5 maps.

Manuscript Received by the Society November 19, 1975
Manuscript Accepted December 2, 1975

Printed in U.S.A.

Geological Society of America
Special Paper 179

Coal deposits in Colombia

Dario Suescún-Gomez
Facultad Nacional de Minas, Apartado Aéreo 14–80, Medellín, Colombia

ABSTRACT

Colombian coal deposits range in age from Maestrichtian-Paleocene in the Sabana de Bogotá and Boyacá basins to middle Oligocene in the Amagá-Titiribí basin in Antioquia. On the basis of age alone, these coal beds should be of lignitic rank, but because of the intense Andean tectonism (especially during Pliocene time) the rank has been increased to bituminous coal with high to medium volatile-matter content. In some areas, as a result of thermometamorphic action by the intrusion of porphyritic andesite, the coal beds have been transformed to meta-anthracite. Chemical analyses of the best known coal deposits show a large range in rank.

The main basins studied are Amagá-Titiribí (middle Oligocene), Santander (Paleocene-Oligocene), Valle del Cauca (Eocene), Cerrejón and Jagua de Ibirico (Paleocene), and Boyacá–Sabana de Bogotá (Maestrichtian-Paleocene).

Evaluations made in Colombia in 1968 were 5 billion metric tons of exploitable coal, of which 5 percent is measured coal, 15 percent is indicated coal, and 80 percent is inferred coal. There are some basins yet to be explored and evaluated; when this is accomplished, total coal resources in Columbia may be as much as 10 billion metric tons.

INTRODUCTION

The coal deposits of Columbia are the most extensive in Latin America. Resources in beds suitable for development probably total more than 10 billion metric tons (hereafter, t = metric tons). Annual production of 3 million t is one of the largest in Latin America. This output is consumed totally in Columbia for production of electrical and mechanical energy, for the manufacture of coke, and for domestic use.

Even though Colombian coal deposits are young, ranging in age from Maestrichtian-Paleocene in the Eastern Cordillera basins to middle Oligocene in the Western Cordillera, they range in rank from lignite with 4,500 cal/g to meta-anthracite with 8,000 cal/g. The intensive tectonism of the Andes and thermometamorphism by many intrusive bodies of late Tertiary age have raised the rank of the coal from lignite to bituminous and locally to meta-anthracite. The bituminous coal is suitable for the manufacture of metallurgical coke.

The main centers of coal production and consumption are close to Colombia's three principal cities: Bogotá with 4 million inhabitants, Medellín with 1 million inhabitants, and Cali with 850,000 inhabitants. Another major center is the great iron-ore smelting center of Belencito, site of the steel works of Paz del Río.

Historically, the policy of mineral exportation in Colombia has been weak. Because of this — and despite the considerable coal resources and the large demand for coal in Latin American markets — there has not been sufficient impetus for industry to develop coal for export. This lethargy is heightened by high mining costs with rudimentary techniques, by large internal transport costs (transportation is hampered by the fact that roads and railroads in the mountains were not built for economic transportation of minerals), by lack of washing plants, and by nonexistent port facilities, both on the Atlantic and Pacific coasts. These factors do not permit Colombian coal to be internationally competitive.

The coal-chemical industries have not progressed in Colombia because there are no chemical plants for by-products. The coke industry serves only the steel factories. Annual production of coke by-products is roughly 15,000 t of coal tar, 3,000 t of ammonium sulfate, 3,000 m^3 of benzol, 60 t of naphthalic compounds, and 600 t of crude tar. Coke for other uses is produced in "beehive furnaces" in which the by-products are discharged into the air and cannot be recovered.

Another important factor that has slowed mineral production in Columbia, and therefore has hindered production of coal, is legislation that has not allowed modernization of present installations nor the application of advanced technology. The Colombian concept of the rights of and obligations to the investors is typically colonial Spanish, a system that for centuries has not permitted the training and employment of enough Colombian engineers and experts to develop the mining sector of the economy; thus this system offers no stimulus for the creation of a national minerals market. Instead, the existing policies actually limit the volume of minerals required by each industry to manufacture even the most necessary products.

COAL BASINS

Colombian coal deposits occur in two major basins that have different geologic features: the eastern basin of miogeosynclinal origin and the western basin of eugeosynclinal origin (Fig. 1).

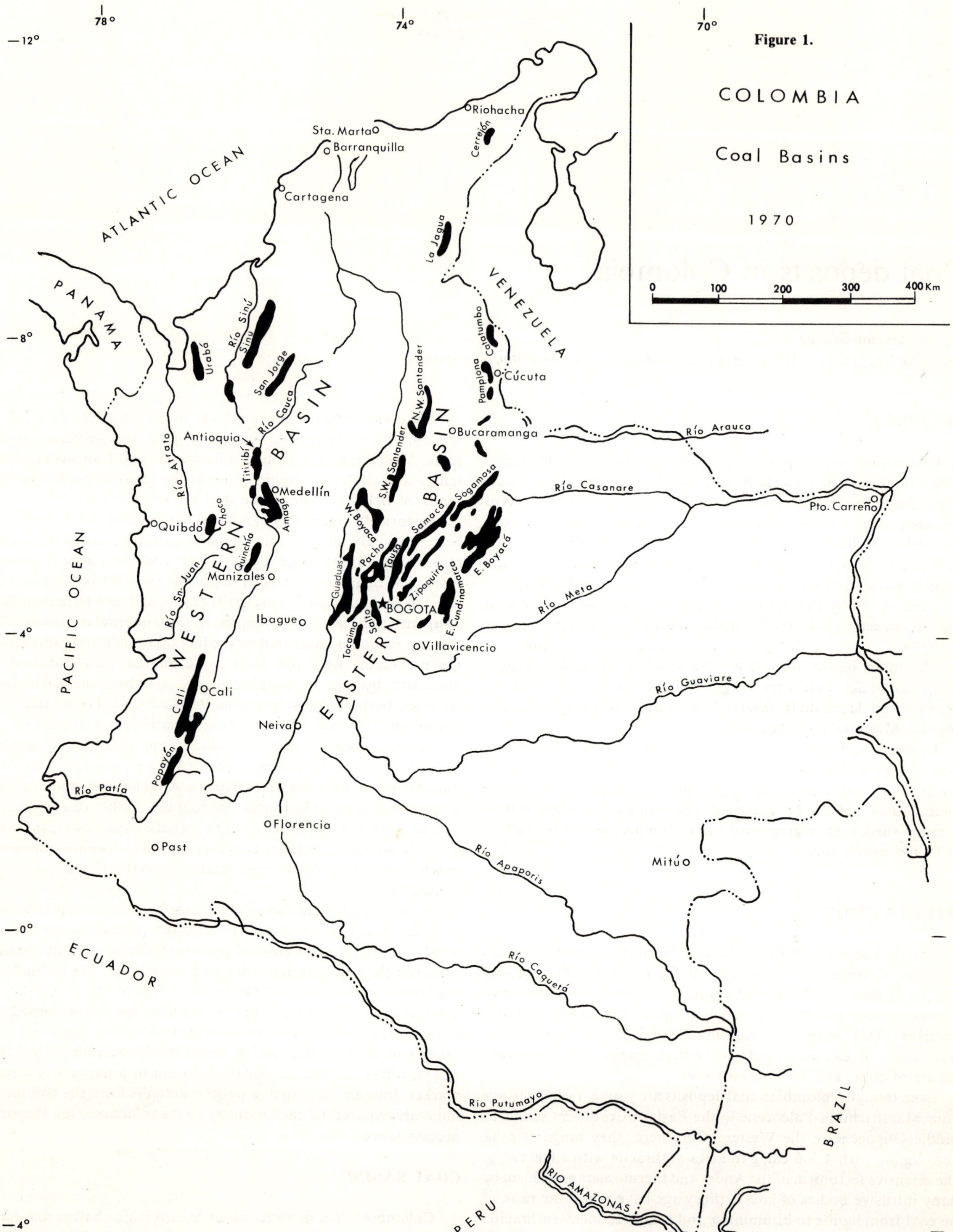
COLOMBIA
Coal Basins
1970
0 100 200 300 400 Km
78°
74°
70°
12°
8°
4°
0°
4°
ATLANTIC OCEAN
PACIFIC OCEAN
PANAMA
VENEZUELA
ECUADOR
PERU
BRAZIL
WESTERN BASIN
EASTERN BASIN
Riohacha
Sta. Marta
Barranquilla
Cartagena
Cerrejón
La Jagua
Catatumbo
Pamplona
Cúcuta
Bucaramanga
N.W. Santander
S.W. Santander
W. Boyacá
Sogamosa
Samacá
Pacho
Tausa
Zipaquirá
E. Boyacá
E. Cundinamarca
Guaduas
BOGOTA
Tocaima
Salto
Villavicencio
Urabá
Río Sinú
Sinú
San Jorge
Río Cauca
Antioquia
Titiribí
Medellín
Amagá
Río Atrato
Quibdó
Chocó
Quinchía
Manizales
Ibague
Río Sn. Juan
Cali
Neiva
Popayán
Río Patía
Past
Florencia
Río Arauca
Río Casanare
Pto. Carreño
Río Meta
Río Guaviare
Mitú
Río Apaporis
Río Caquetá
Río Putumayo
RIO AMAZONAS
Figure 1.

Eastern Basin

The coal deposits in the eastern miogeosynclinal basin of the Eastern Cordillera were formed during the Cretaceous-Tertiary transition period, mostly during Maestrichtian, Paleocene, and Eocene time. The best coal is associated with the Guaduas Formation of early Tertiary age.

The eastern basin is subdivided into the following subbasins: Tocaima-Guaduas, El Salto–Pacho, Bogotá-Zipaquirá-Villapinzón, eastern subbasin of Cundinamarca, Samacá-Tunja, Sogamoso-Tópaga, eastern subbasin of Boyacá, western subbasin of Boyacá, southwestern subbasin of Santander, northwestern subbasin of Santander, Pamplona-Cúcuta-Catatumbo, Jagua de Ibirico, and El Cerrejón.

Western Basin

The coal deposits in the western eugeosynclinal basin, which is west of the Magdalena River between the Central and Western Cordilleras, are the youngest and also the most highly metamorphosed coals in Columbia. They range in age from Eocene-Oligocene to Miocene. The younger coal is markedly lignitic; only Miocene-Oligocene and older coals can be classified as bituminous.

The western basin is subdivided into the following subbasins: Popayán-Cali, Quinchía–Río Sucio, Amagá-Titiribí, Río San Jorge, Río Sinú, and Urabá.

Estimated Coal Resources

In Colombia, the coal industry has been developed only for local consumers, and most installations are very rudimentary; only four mining operations are semimechanized. Studies of resources are restricted to a few deposits and give only a slight indication of the total potential of the country. Table 1 shows the estimated coal resources of Colombia.

Deposits in the Eastern Basin

Not all of the subbasins known in the eastern basin have been developed or studied in sufficient detail. This discussion is restricted to those subbasins that are in production or that may be fundamental for the future economic development of Columbia.

Tocaima-Guaduas Subbasin

The Tocaima-Guaduas subbasin is a large, north-striking syncline that passes near the town of Guaduas, 100 km west of Bogotá. This large syncline formed during early Paleocene time and shows a good concentration of coal in a belt 150 km long and 8 km wide. Few detailed studies have been carried out in this area because tectonism has crushed the vitreous coal into small pieces, which on mining yield 40 percent fines. This reduces the value and utility of the mined product; however, the fines could be burned in a cyclone furnace for the generation of electric power.

The geologic importance of this subbasin is that near Guaduas the Colombian type section of the lower Paleocene rocks was studied and named the Guaduas Formation. This was the starting point for the geologic study of the other coal-bearing basins in eastern Colombia.

TABLE 1. ESTIMATED COAL RESOURCES OF COLOMBIA

Resource category	1968 Reserves	
	(t)	(%)
Measured	250,000,000	2.5
Indicated	750,000,000	7.5
Inferred	4,000,000,000	40.0
Possible additional	5,000,000,000	50.0
Total	10,000,000,000	100.0

Bogotá Area Subbasins

In the area of Bogotá, the Guaduas Formation crops out in a wide area. Its principal structures are oriented N30° E and range from simple folds to complicated forms with large amounts of thrust faulting. The Guaduas Formation consists of brown clay, carbonaceous shale, yellow sandstone, red and violet lutite, and beds of mineable coal.

The Guaduas Formation has been subdivided into lower, middle, and upper divisions. In the lower Guaduas, sandy sedimentary rocks predominate and layers of coal are scarce. Coal layers are, however, low in volatile matter (17 to 20 percent), which allows them to be mixed with coal of higher volatile content to yield the best metallurgical coke in Colombia.

The middle Guaduas (600 m thick) is characterized by thick beds of shale that are subdivided by the La Guía Sandstone and the the La Lajosá Sandstone and by good concentrations of mineable coal beds, which make this unit the best coal producer.

The upper Guaduas is formed by a sequence of red shale between the La Lajosa Sandstone and the El Cacho Sandstone. Coaly beds are absent. Some thin lignite beds are the only carbonaceous strata.

Palynological studies by the Servicio Geológico de Colombia have established a Maestrichtian age for the lower and middle Guaduas and a Maestrichtian-Paleocene age for the upper Guaduas.

The subbasins of the Bogotá area include the Suesca-Chocontá zone, with the coal beds of the Ovejeros and Cogontá anticlines and the Santa Rosita syncline. The Lenguazaque syncline and the Sutatausa anticline belong to the coal-bearing zone of Tausa-Guachetá. The coal deposits of San Vicente, Santa Rosita, Manacá, and Cogontá belong to the Suesca-Chocontá zone, which contains seven beds that range from 2.5 to 0.8 m thick. These are the main coal-producing beds of the zone.

The coal deposits of Cogua, Tausa, Sutatausa, Cucunubá, Lenguazaque, and Guachetá are in the zone of Tausa-Guachetá. These deposits are mined, not according to the size, shape, and disposition of the coal bed or beds, but according to the numerous small mining concessions granted arbitrarily by the government without regard to coal-bed continuity.

The petrographic composition of the coals of the Bogotá subbasins averages clarain, 80 percent; vitrain, 9 percent; durain, 8 percent; and fusain, 3 percent.

The chemical composition of the coals differs appreciably from one basin to another, as well as within a single basin. In general, the coals increase in rank with increase in geologic age. The general composition of the main coal zones is shown in Table 2. The reserves measured in the zones are summarized in Table 3.

Tunja Area Subbasins

The Tunja area subbasins include the Guaduas Formation coals in the Department of Boyacá. These subbasins present geologic and mineralogic features similar to those in the Bogotá subbasins. The major production is from the Samacá mines, which are in the Guachetá-Samacá subbasin, and from the Chapa mine in the Sogamoso-Tópaga subbasin. From the Chapa mine, the steel factory of Paz del Río obtains coal of low-volatile bituminous rank for use in blends to obtain coke suitable for use in smelting iron ore. This plant produces 1,000 t of coke daily.

TABLE 2. GENERAL COMPOSITION OF COALS OF BOGOTÁ AREA SUBBASINS

Zones	Volatile matter (%)	Fixed carbon (%)
Tausa	25.33–37.51	68.63–55.66
Suesca	32.41–38.21	48.14–58.46
Guachetá	17.82–29.60	78.22–65.08

TABLE 3. COAL RESERVES OF BOGOTA AREA SUBBASINS

Zones	Total coal thickness (m)	Reserves (t)
Tausa-Cucunubá	17.71	2,092,000,000
Lenguazaque-Cucunubá	11.57	2,068,000,000
San Vicente	6.92	1,300,000,000
Santa Rosita–Manaca	8.92	751,000,000
Cogontá	7.90	70,000,000

Measured resources are 18 million t in the eight deposits being developed; 138 million t of indicated reserves occur in these subbasins.

The average chemical analysis of the coals of this zone is moisture, 1 to 4 percent; volatile matter, 18 to 42 percent; sulfur, 0.4 to 1.6 percent; ash, 3 to 15 percent; calorific value, 12,000 to 15,000 Btu/lb.

Santander Subbasin

The Santander subbasin is in the region of Landázuri; the coal zone begins 1 km north of the town of Landázuri and continues to the "Vuelta de Baticola." The zone is more than 1 km wide and contains as much as 12 m of mineable coal in beds as thick as 2 m.

The Landázuri coal is high in calorific value and low in volatile

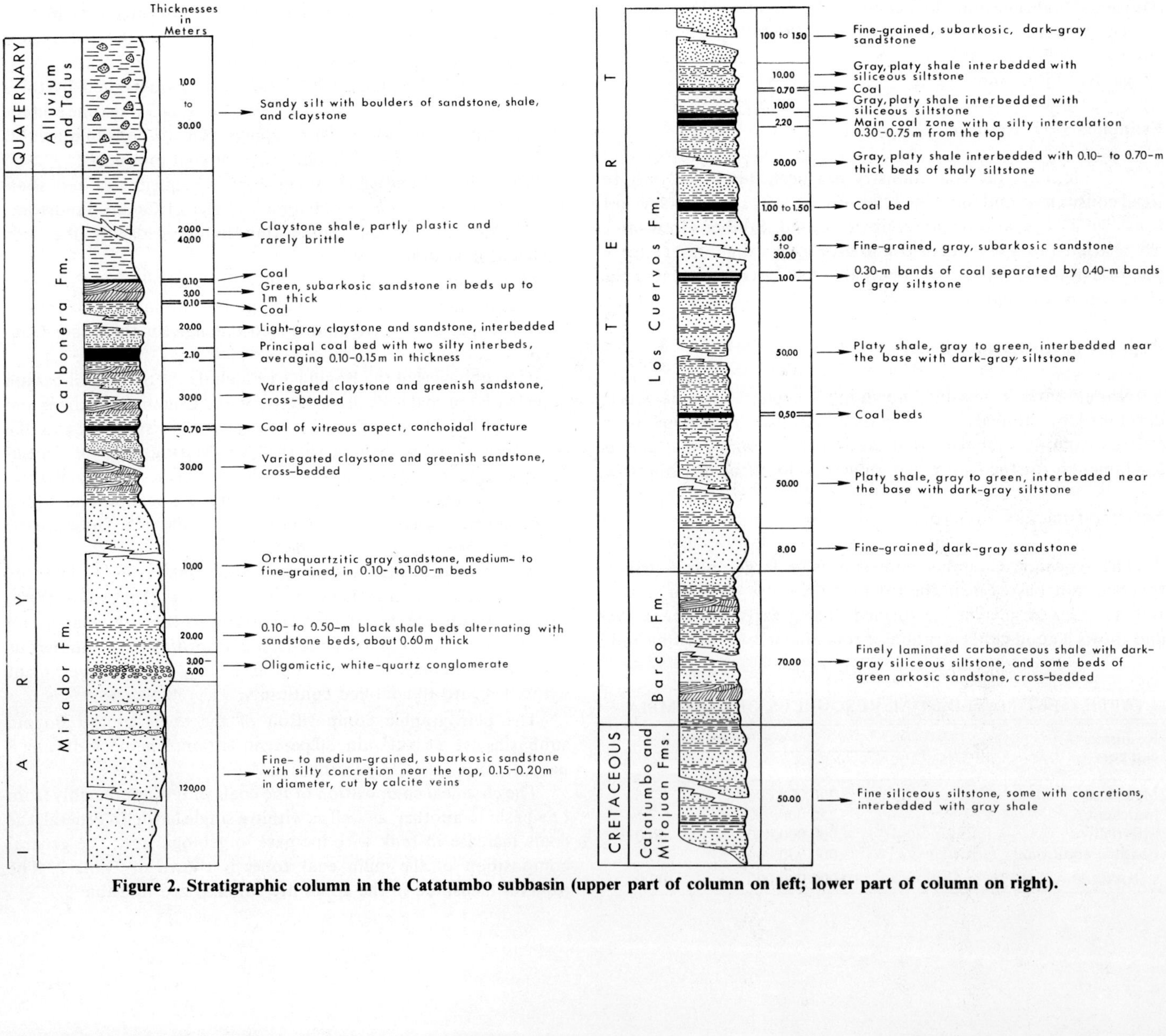

Figure 2. Stratigraphic column in the Catatumbo subbasin (upper part of column on left; lower part of column on right).

matter but is not of coking quality. It is difficult to burn. A representative chemical analysis is moisture, 1.0 to 2.7 percent; volatile matter, 8.8 to 17.9 percent; fixed carbon, 74.1 to 84.1 percent; ash, 2 to 7 percent; and calorific value, 7,000 to 8,000 cal/g.

A concession for studying and developing the coal has been granted by the Colombian Government to a Colombian-Spanish firm. The mining development began with a plan to export 2 million t annually of anthracite. The coal would be hauled by trucks on a road 105 km long to the Magdalena River and from there would be shipped on large barges to the port of Cartagena on the Atlantic coast for export to Spain.

Pamplona-Cúcuta-Catatumbo Subbasin

The Pamplona-Cúcuta-Catatumbo subbasin (Fig. 2) is near Venezuela, a fact which aids in the formation of integrated economic cooperative projects between the two countries. Among other positive features of this coal basin are high-quality coking coals in various beds.

In the Maracaibo basin, the lower Oligocene Carbonera Formation and the middle Eocene Los Cuervos Formation are well defined. Lignitic and bituminous coal beds are present in the Carbonera Formation. The bituminous coals are rated at greater than 6,000 cal/g. The thickness of the beds ranges from 0.1 to 2.2 m, but only four beds (with thicknesses of 0.8, 1.0, 1.2, and 2.2 m, respectively) are mineable. Most of these beds have good coke indexes. Present measurements show 15 million t of mineable coal.

The Los Cuervos Formation consists principally of sandy shale, red variegated shale, and as many as 16 thin beds of coal. Only five coal beds are exploitable because they are more than 0.8 m thick. The presence of many thin coal beds in the Paleocene layer indicates the continental nature of the depositional environment. The Los Cuervos Formation occurs as part of a large eastward-dipping monocline. The unit overlies the igneous-metamorphic Durania basement. The mineable coals are in the middle and lower parts of the Los Cuervos; dips range from horizontal to vertical.

An average chemical analysis shows moisture, 3 percent; volatile matter, 30 percent; fixed carbon, 60 percent; ash, 7 percent; sulfur, 1 percent; and calorific value, 6,000 cal/g or more.

Jagua de Ibirico Subbasin

In the middle-eastern region of the Department of César, on the eastern piedmont of the Perijá Range, Paleocene-Eocene strata are exposed near the town of Jagua de Ibirico in a slightly folded monocline that has a general eastward dip of 30° and is in fault contact with the underlying crystalline basement of the Perijá Range.

Black slaty shale, yellow sandstone, and black shale predominate in the section. In a belt 10 km long from north to south and 100 m in stratigraphic thickness, beds containing more than 100 million t of high-volatile bituminous coal have been found. The coal should be relatively easy to mine and to market. Detailed studies have not been made, but reconnaissance work has shown the presence of coal beds as much as 5 m thick. Typical analyses are moisture, 2 to 10 percent; volatile matter, 32 to 47 percent; fixed carbon, 48 to 58 percent; ash, 1 to 5 percent; sulfur, 0.5 to 1 percent; and calorific value, 11,500 to 13,600 Btu/lb.

Even though the coal is not coking coal, the basin has several favorable characteristics such as thick coal beds, low ash content of the coal, and good transportation facilities, being only 30 km from the Atlantic coast railroad line and 260 km from the seaport of Santa Marta.

El Cerrejón Subbasin

The coal in the Cerrejón subbasin ranges from metamorphosed lignitic to bituminous coal high in volatile matter. It is not coking coal but is used in blends with other Colombian coals to produce acceptable metallurgical coke.

This subbasin is in the northern Department of Goajira. The coal occurs in a subaqueous deltaic facies characterized by lenticular sandstone and laminated shale, with gradual transitions between the fine-grained and coarse-grained sedimentary rocks. The thickness of the sequence ranges from 900 to 1,000 m; it contains as many as 22 beds of coal interbedded with shale and sandstone. The mineable beds range in thickness from 2 to 6 m, but a few beds are thicker than 10 m. In general, the coal deposits form a monocline dipping 30° E. The upper part of the sequence (with dips of 10° E) contains as much as 14 m of coal with a surface cover of only 30 m. Extraction of the coal in open-pit strip mines can be done in bands not less than 500 m wide.

The Cerrejón coal basin is the best studied in Colombia. The data are on file at the Instituto de Fomento Industrial de Bogotá. Measured reserves are 32 million t and indicated reserves are 312 million t.

The average quality of coal in the Cerrejón area is volatile matter, 38.3 to 43.2 percent; fixed carbon, 52.6 to 59.2 percent; ash, 1.7 to 5.2 percent; sulfur, 0.5 to 1.0 percent; calorific value, 13,500 to 13,890 Btu/lb; and ash fusion point, 2,230° to 2,570° F.

The center of this subbasin is 100 km from the Atlantic coast, near the seaport of Riohacha.

DEPOSITS IN THE WESTERN BASIN

The coals in the western basin of Colombia are of Eocene and Miocene-Oligocene ages. The soft coals in most places are intruded and thermometamorphosed in the zone of Pliocene neovolcanism, corresponding in time with the final paroxysm of the Andean orogeny.

Popayán-Cali Subbasin

The Cenozoic coal deposits in the Popayán-Cali subbasin (Fig. 3) extend from the latitude of Popayán, along the Río Cauca valley, to north of Cali. They are included in a narrow belt of continental sedimentary rocks about 1,500 m thick. Conglomerate and sandstone predominate in the lower part; claystone and sandstone alternating with coal beds constitute the middle part; and shale and coal beds too thin for economic development form the upper part. This formation is possibly of Oligocene age according to palynologic studies made by the Servicio Geológico de Colombia. It lies discordantly over the "Diabasic group" of the Upper Cretaceous strata. In the Popayán area, many igneous intrusions produced the anthracitic character of the coal. In the Cali area, the coal is high in volatile matter, with some beds of coking coal. However, the structure is so complicated that about 30 percent of the coal is fines (ciscos) that are not usable without agglomeration into briquettes.

The best known and best developed deposits in this subbasin are the San Francisco, Timba, La Bohemia, Pance, Río Claro, and Golondrinas. Calculated resources have been estimated from a high of 400 million t to a low of 16 million t of strippable coal. The most reasonable evaluation is 60 million t of mineable bituminous and higher rank coal (Suescún, 1959).

The ranges of analyses for anthracitic coal in the south zone of the Popayán area are as follows: fixed carbon, 68.67 to 84.12 percent; volatile matter, 8.25 to 14.00 percent; ash, 4.34 to 21.99 percent; sulfur, 0.69 to 4.59 percent; calorific value, 6,258 to 8,344 cal/g or 11,264 to 15,018 Btu/lb. Coke button; agglutinating properties, and hardness are fair to good. Bituminous coal in the north zone of the Cali area are as follows: fixed carbon, 34.65 to 58.12 percent; volatile matter, 31.17 to 44.12 percent; ash, 4.44 to 24.60 percent; sulfur, 0.33 to 5.92 percent; calorific value, 6,149 to 7,942 cal/g or 11,067 to 14,331 Btu/lb. Coke button; agglutinating properties, and hardness are fair to good.

In Cali, a washing plant is managed by the Instituto de Fomento Industrial. It has a washing capacity of 120 t/hr, but it has not been used to date because the coal could not be exported through the Pacific seaports owing to the high cost of transportation.

Antioquia Subbasin

An excellent geologic report was published in 1926 by Emil Grosse in which he described in great detail the Miocene-Oligocene stratigraphy in this subbasin (Fig. 4) and calculated the coal resources. This work by Grosse is a model of professional work and clarity of interpretation which has served as an example for the new generations of Colombian geologists.

The lower part of the coal formation is 200 m thick. It is characterized by the presence of light-colored conglomerate, white

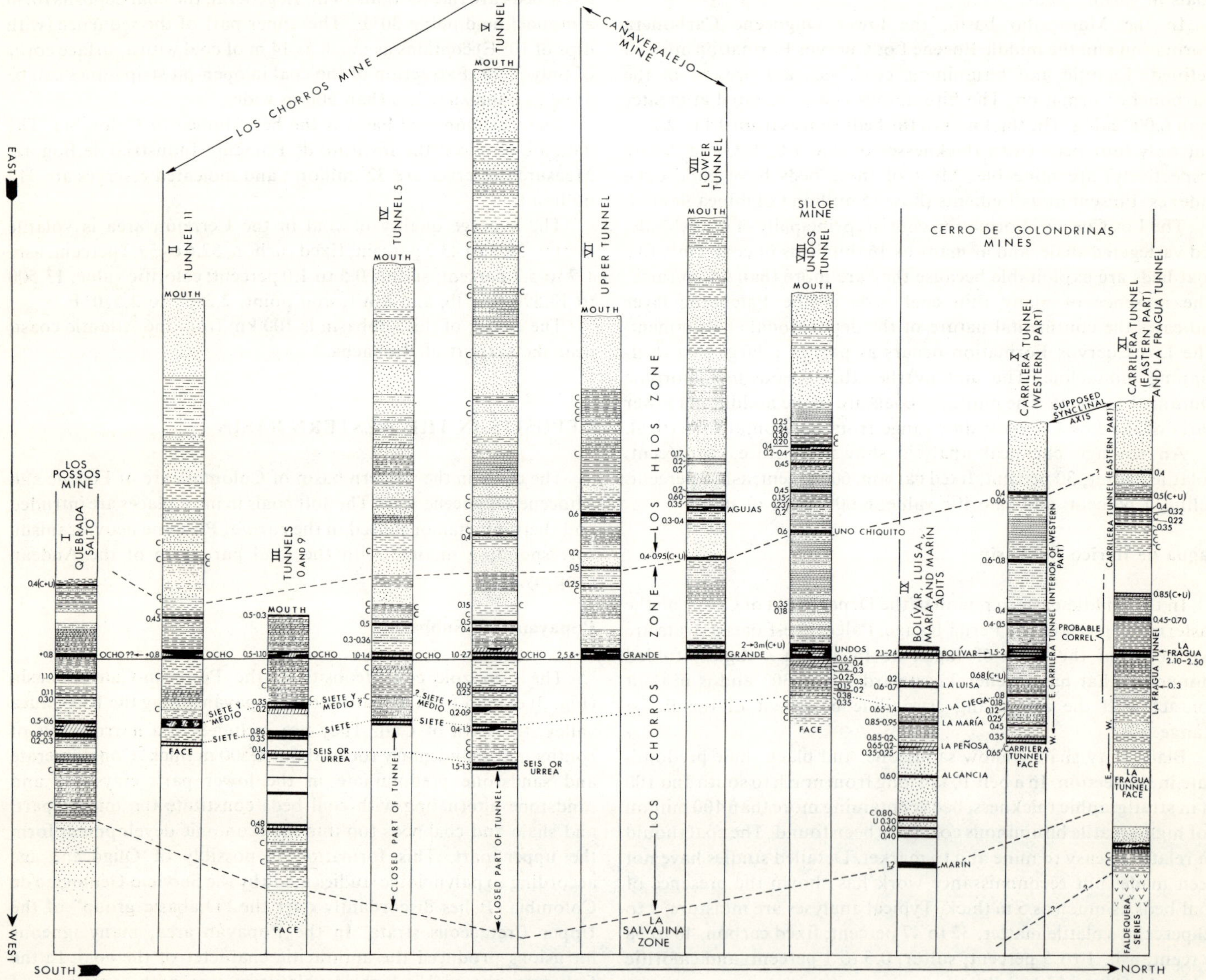

Figure 3. Stratigraphic column in the Cali subbasin.

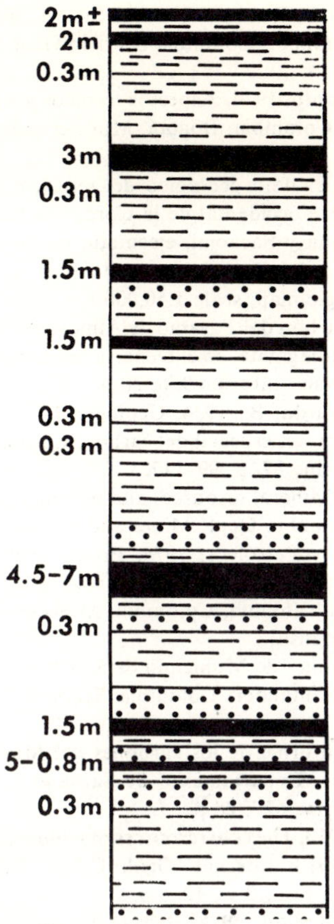

Figure 4. Stratigraphic column in the Antioquia subbasin (Titiribi).

sandstone, and gray slaty shale with thin coal beds no more than 0.3 m thick.

The middle unit, which contains the productive coals, has a thickness of 250 m. The sedimentary rocks are predominately gray slaty shale and white sandstone with a notable absence of conglomerate. Coal beds are present in mineable thicknesses of as much as 10 m. Most of the coal beds range in thickness from 0.75 to 2.85 m.

The upper unit contains no mineable coal, lacks conglomerate, and is mainly of gray to green sandstone that, because of intensive weathering, exhibits a brick-red color in surface outcrops. The total thickness of the unit is about 300 m.

The basement of this subbasin is an igneous-metamorphic complex consisting of Paleozoic phyllite to gneiss, amphibolite of undetermined age, and Mesozoic porphyritic bodies. The coal formation is folded, faulted, and intruded by many late Tertiary igneous rocks.

Grosse (1926) estimated the coal resources of this subbasin to total 1,500 million t, including outcropping coal, near-surface coal, and coal in the anticlinal and synclinal structures. In 1969, various deposits were being mined in this subbasin, the more important being near the towns of Fredonia, Venecia, Titiribí, Amagá, and Angelópolis, all in the Department of Antioquia (Fig. 4).

The range of chemical analyses shows fixed carbon, 41.92 to 59.20 percent; volatile matter, 29.70 to 47.94 percent; moisture, 3.65 to 11.37 percent; ash, 1.29 to 6.20 percent; sulfur, 0.40 to 1.20 percent; and calorific value, 5,071 to 6,271 cal/g.

This coal formation continues into the north and northeastern parts of the Department of Antioquia in the regions of Tarazá, Río San Jorge, Río Man, and Urabá. Exploration of these subbasins has been very restricted, and the few outcrops studied show that they contain lignitic coal in thin beds.

CONCLUSIONS

Colombia has the largest coal resources of the Latin American countries. Potential resources are 10 billion t.

Colombia is one of the largest producers of coal in Latin America but does not export coal. The annual output of 3 million t is used internally.

The Colombian coal beds, with the exception of a part of the Cerrejón deposit in Goajira, have to be mined by underground methods in areas of rather complex geologic structures.

An integrated geologic study and an exploration program for Colombian coals are necessary to understand the occurrence and distribution of beds and to discover new deposits.

Colombia is a developing country. To create employment, it obviously should prefer to use its own coal and other natural resources before importing foreign products.

Although there are coal beds in Colombia with a very high percentage of ash, as in the Cauca Valley coals, even these coals can be used to produce electrical energy or to produce steam for various industries such as paper, cardboard, and sugar refineries.

The majority of Colombian coals do not yield coke of good quality because there is a lack of local agglutinating coals. Colombian coals can be blended with coals from Chile, Argentina, and Brazil in mixtures of 30 percent Colombian coal and 70 percent imported coal to produce excellent coke, according to studies carried out by the Instituto de Fomento Industrial. With a more advanced technology, a metallurgical coke of good specifications could be obtained from Colombian coals. Unfortunately, in Colombia the technique of mixing coking coals, the process of compacting coals to 950 kg/m^3, and the drying of coals before processing in the coke ovens are not well known.

More advanced scientific and technological investigations will discover more uses for coal fines (perhaps as antifissurants in the making of coke), semicoke, low-grade coal, and anthracite. This may require foreign help.

In Colombia, the use of coal in thermoelectric plants is hampered by the lack of washing plants and by the potentially large residue of ash. In the future, the ash residue may be used for incorporation in commercial cements to take advantage of its pozzolanic qualities; for use in road building and in reinforced concrete, for use in building structures that are resistant to water corrosion, for replacing the sand in cellular concrete, and for manufacturing light aggregates.

SELECTED REFERENCES

Grosse, Emil, 1926, Das Kohlentertiar Antioquias — El Terciario carbonífero de Antioquia: Berlin, Verlag Von Dietrich Reimer (Ernst Vohsen), 361 p.

Hincapie, Stella, 1971, La industria del carbón en Colombia: Bogotá, Ministerio de Minas y Petróleos, Oficina de Planeación, 117 p.

Molina, Elkin, 1965, El Cretáceo carbonífero al norte de la Sabana de Bogotá; Las cuencas carboníferes de Boyacá y Cundinamarca: Bogotá, Bol. Geológico, v. 13, 99 p.

Suescún-Gomez, Darío, 1959, Plan decenal de desarrollo, *in* Chapter on transporte y energía: Bogotá, Dept. Nacional Planeación, 72 p.

OTHER SOURCES

Instituto de Fomento Industrial (IFI), Unpub. reports in the IFI library on coal deposit of various basins in Colombia.

Colombian Mining Congresses, I, II, III, and IV: Bogotá, Memoirs from 1970 to present.

Manuscript Received by the Society March 12, 1975
Manuscript Accepted March 17, 1975

SUPPLEMENTARY BIBLIOGRAPHY ON COAL IN COLOMBIA

E. R. Landis and Paul Averitt, Compilers
U.S. Geological Survey, Federal Center, Denver, Colorado 80225

The following is a selected bibliography with emphasis on reports in English. For a more complete bibliography, see Beltrán (1974).

Beltrán Cortés, Ernesto, ed., 1974, Carbones de Colombia: Bogotá, Colombia, Ministerio de Minas y Petróleos, 254 p. (Includes a 627-entry bibliography.)

Del Río, A., 1947, Carbones de los Municipios de Pamplonita, Bochalena, y Cornejo, Departamento de Norte de Santander: Colombia Servicio Geol. Nac., Informe 540, 27 p.

Dewey, L. S., 1974, The mineral industry of Colombia, *in* Area reports—International: U.S. Bur. Mines Minerals Yearbook, 1972, v. 3, p. 237–241. (See also prior and subsequent volumes in this series.)

Economic Commission for Latin America, 1957, Energy in Latin America: United Nations, E/CN.12/384/Rev. 1.

El Cerrejón coal field: Bogotá, Colombia, Inst. Fomento Industriales Colombia, 38 p.

Feys, R., Geffroy, J., and Vetter, P., 1967, Un cas nouveau de graphitisation du charbon par une intrusion éruptive a Cali (Colombia): Geol. Soc. Nord Ann., v. 87, no. 3, p. 145–150.

Fraser, Thomas, 1948, Coal in South America: Pan Am. Inst. Mining Engineering and Geology, U.S. Sec. Tech. Paper 6, 16 p.

García, C., and Manjarres, L. T., 1962, Investigación carbonífera de la zona comprendida entre Ricaurte-Lourdes-Silos-Labateca, Departmento de Norte de Santander: Colombia Servicio Geol. Nac., Informe 1417, 83 p.

García Jacome, Eduardo, 1974, Aspectos técnicos y legales de las cuencas carboníferas en Colombia: Bogotá, Ministerio de Minas y Petróleos, 67 p.

Grosse, Emil, 1926, Estudio geológico del terciario carbonífero de Antioquia en la parte occidental de la Cordillera Central de Colombia, entre el Río Arma y Sacaojal, ejecutado en los años de 1920–1923, para el gobierno del Departamento de Antioquia: Berlin, Reimer. (Contains geologic map, scale 1:50,000.)

Hubach, E., 1953, Carbón del Páramo de Almorzadero y carbón y fuentes de agua sal de Molagavita, en el Departamento de Santander: Colombia Inst. Geol. Nac., Informe 926, 20 p.

Hubach, E., and Alvardo, B., 1934, Geología de los Departamentos del Valle y del Cauca, en especial del carbón, Colombia; Colombia Inst. Geol. Nac., Informe 87, p. 99–117.

Jimenez, J. J., 1933, Mineral resources and mining regulations of Colombia: Bogotá, Colombia Dept. Mines and Petroleum, pt. 1, 26 p.

Ministerio de Minas y Petróleos, 1974, Primer seminario internacional sobre la utilización integral del carbón, Bogotá, Colombia, March 10–17, 1974 [abs.]: Bogotá, Colombia, Ministerio de Minas y Petroleos, 40 p. [in English].

Paba, F., 1946, Carbones de Municipio de San Vincente de Chucuri, Departamento de Santander: Colombia Servicio Geol. Nac., Informe 525, 8 p.

——1948, Segundo informe de los carbones del Municipio de San Vincente de Chucuri, Departamento de Santander: Colombia Servicio Geol. Nac., Informe 627, 28 p.

Pereira Gamba, F., 1913, The coal resources of Colombia, *in* Coal resources of the world, 1913: Internat. Geol. Cong., 12th, Toronto, 1913, Morang and Co., Ltd., 3 v., 1266 p.

Reichenbach, Richard Paul, 1928, Beitrag zur kenntnis der Kohlen der kolumbianischen Ostkordillere (Kordillere von Bogotá): Berlin, im Vertrieb bei der Preussischen Geologischen Landesanstalt.

Singewald, Q. D., 1950, Mineral resources of Colombia: U.S. Geol. Survey Bull. 964-B, 204 p. (Note p. 94–109.)

Suescún-Gómez, Dario, 1969, Coal deposits in Colombia: Geol. Soc. America, Abs. for 1968 Ann. Mtg., Spec. Paper 121, p. 114–115.

U.S. Bureau of Mines, 1961, 1971, 1972, 1973, 1975, International coal trade: Washington, D.C., U.S. Bur. Mines, v. 30, no. 10, p. 24–25; v. 40, no. 11, p. 7–8; v. 41, no. 10, p. 7; v. 42, no. 12, p. 5–6; v. 44, no. 6, p. 9.

Ward, Dwight E., Goldsmith, Richard, Cruz B., Jaime, and Restrepo A., Hernán, 1970, Mineral resources of the southern half of zone III, Santander, Norte de Santander, and Boyaca, Colombia: U.S. Geol. Survey Open-File Rept., (IR) CO-17, 160 p.

Wokittel, R., 1953, Yacimientos del Municipio de Molagavita, Departamento de Santander: Colombia Servicio Geol. Nac., Informe 972, 21 p.

Manuscript Received by the Society September 9, 1975
Manuscript Accepted December 2, 1975

Printed in U.S.A.

Geological Society of America
Special Paper 179

Coal deposits of Latin America

Wilds W. Olive
U.S. Geological Survey, Lexington, Kentucky 40503

ABSTRACT

Commercial coal deposits ranging in rank from lignite to anthracite are present in Argentina, Brazil, Chile, Colombia, Mexico, Peru, and Venezuela. Some of the bituminous coal is weakly coking. The commercial deposits are of Triassic, Jurassic, Cretaceous, and Tertiary ages, except those in Brazil, which are of Pennsylvanian and Permian ages. Most coal beds are less than 3 m thick, but some are as much as 5 to 8 m thick. Much of the coal in Brazil and Mexico occurs in flat-lying or gently dipping rocks. In the other countries, the coal beds commonly are folded and faulted and at places are intruded by igneous rocks associated with the Andean uplift.

Rough estimates based on available data indicate that the largest coal resources, about 10 billion metric tons, are in Columbia. Resources in Brazil, Venezuela, Mexico, and Chile are estimated at more than 1 billion metric tons for each country, and in Argentina and Peru, resources amount to more than 250 million metric tons for each country. Detailed geologic investigations are needed to provide a basis for more meaningful estimates.

During 1973, Brazil, Chile, Colombia, and Mexico each produced more than 3 million metric tons, dominantly of bituminous coal. Imported coal consisted almost entirely of low-volatile bituminous coal, which was blended with domestic coal to manufacture coke. Very little coal was exported.

INTRODUCTION

This paper on coal deposits of Latin American countries is largely a synopsis of published reports and of unpublished information from the files of the U.S. Geological Survey and the U.S. Bureau of Mines. The main content is a general description of coal deposits for the five major coal regions of Latin America. Most of the reports have been based on exploration or reconnaissance surveys; very few are based on detailed investigations. As the geology of Latin American coal deposits becomes better known, information undoubtedly will be revealed to warrant revision of much of what is presented here.

As shown in Figure 1, areas of the principal coal deposits of Latin America are widespread. The deposits occur from the northern border of Mexico to the southern tip of South America. The topography and climate of the coalfields are diverse. In some, the relief is gentle, and in others it is moderate to rugged and mountainous. Climate differs with latitude and altitude and with location in relation to ocean currents and to the major wind belts of the world. The climate of areas near the northern coast of South America and in most parts of Mexico is tropical rainy to dry. In the middle latitudes of South America, the climate is humid mesothermal, except near the southern tip where it is marginal between humid mesothermal and polar.

Many coal deposits are near and easily accessible to markets. Other deposits, although near markets, are relatively inaccessible because of ruggedness of terrain and poor means of transportation. Still others, although accessible to good means of transportation, are remote from markets.

The coal measures consist dominantly of continental, fine to coarse clastic sedimentary rocks of Paleozoic, Mesozoic, and Cenozoic ages. In Mexico and Central America, the coal-bearing sequences were deposited in lacustrine and palustral environments in the Mexican and Gulf Coast geosynclines. In South America the deposits accumulated at various places along the Andean trough and in isolated basins peripheral to the Guyana and Brazilan shields.

In the mountainous areas of Mexico and western South America, the coal beds commonly are folded and faulted, and at places they have been metamorphosed. In most other areas, the beds are flat lying or gently dipping.

The rank of the coal ranges from lignite to anthracite, but most is bituminous and subbituminous. The quality differs within wide limits. Some bituminous coal produces a good quality coke, but most is noncoking to weakly coking.

Total reserve estimates for the principal Latin American coalfields are more than 20 billion metric tons (hereafter, t = metric tons) of coal of all ranks. Because of the paucity of data on quality, thickness, extent, and depth of coal beds in many coalfields, the significance of this figure relative to recoverable reserves is uncertain. As a general rule of thumb, the less that is known about a coal deposit, the greater is the tendency to overestimate its reserves and its economic potential.

Although coal is present in many parts of Latin America, the principal deposits are in five major regions. Seven countries are involved: Mexico, Colombia and Venezuela, Peru, Argentina and Chile, and Brazil.

Figure 1. Areas of principal coal deposits of Latin America.

Possibly a sixth region is on the islands and coastal areas of the Chiriqui Lagoon on the Caribbean coast of Panama (Fig. 1). Thick coal beds were reported (Evans, 1861) in that region, but information is inadequate to evaluate their potential. Jackson (1861) considered the deposits to be of late Eocene age, but as shown by the North American Geologic Map Committee (1965), the oldest rocks — presumably those that include the coal-bearing sequence in the area of the lagoon — are now believed to be of Miocene age. According to Evans (1861), the sequence contains as many as 8 coal beds aggregating 10 m in thickness, 8 m of which could be mined together. In places the coal beds dip 15° to 20°. Analyses indicate that the ash and sulfur contents are low and that some of the coal may be suitable for metallurgical coke production. These deposits, which are favorably situated relative to ocean-navigation routes, may have economic potential and warrant further investigation.

MEXICO

As shown by Gonzáles Reyna (1956, map in pocket), coal deposits in Mexico are known in 22 of the 29 states, but little is known about the geology of most deposits. A few in the states of Coahuila, Sonora, and Oaxaca have been investigated to a limited extent. Investigation of others has consisted mainly of examination of outcrops and a few exploratory drill holes.

As described by Gonzáles Reyna (1956, p. 443), the coal

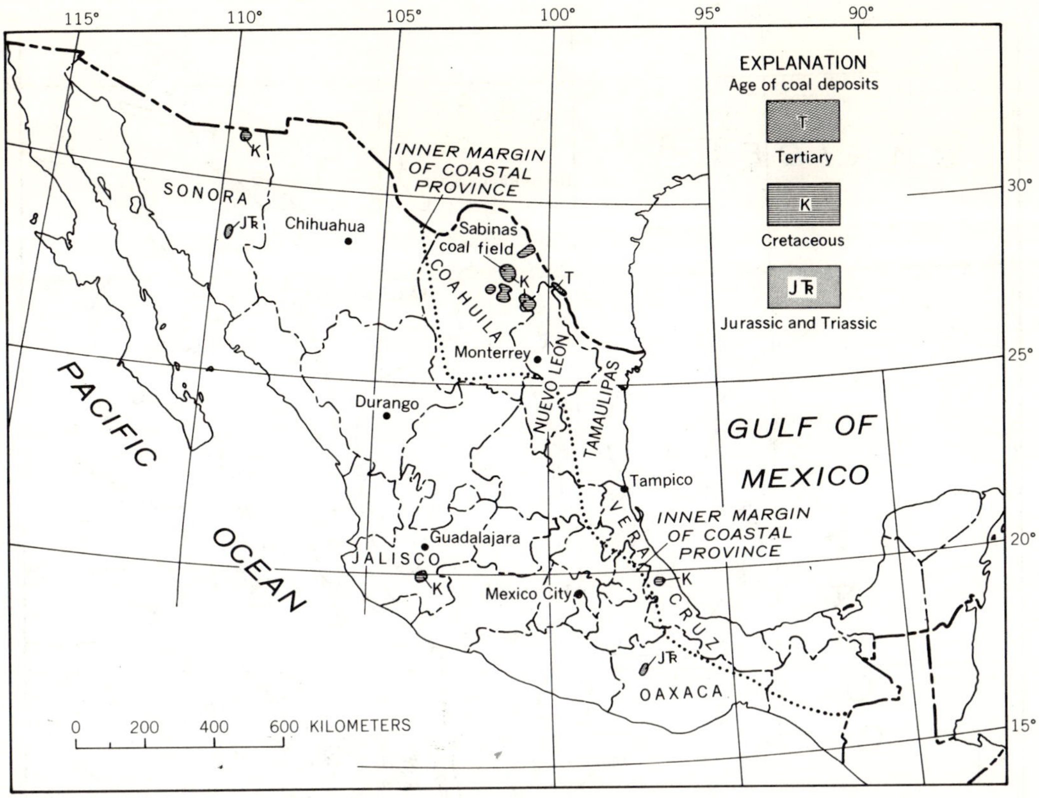

Figure 2. Principal coalfields of Mexico.

measures in most of the principal coalfields, shown in Figure 2, are of Cretaceous age. Coal of Triassic and Jurassic ages occurs in the states of Oaxaca and Sonora (Jesús Ojeda-Rivera, 1968, oral commun.), and deposits of Tertiary age border the Rio Grande in the states of Nuevo León and Tamaulipas.

Cretaceous coal deposits in northeastern Mexico, because they are the source of nearly all of Mexico's coal production, have been studied in greatest detail. These deposits are in several separate basins, most of which are in the state of Coahuila. According to the U.S. Bureau of Mines (unpub. data), coal beds ranging in thickness from a few centimetres to as much as 5 m have been found at various levels to depths of more than 1,000 m. The average thickness of coal beds in mines ranges from 1 to 1.5 m. The coal beds in most of these basins are gently dipping, but in some the beds are inclined 10° to 30°; locally, they dip as much as 80°. At places, the coal beds are offset by faults.

In other parts of Mexico, the coal beds commonly are folded and faulted. In Oaxaca and Sonora, they have been metamorphosed, presumably by igneous intrusions.

The rank of the coal ranges from lignite to anthracite; most is bituminous and contains large percentages of ash and small to moderate amounts of sulfur. Both ash and sulfur contents are reduced considerably by washing. After washing, some of the low-volatile bituminous coal, particularly that from the Sabinas coalfield in Coahuila, is suitable for production of metallurgical coke.

Recoverable reserves of bituminous coal and lignite in Coahuila have been estimated by Ojeda-Rivera (this volume) at 3.173 billion t.

COLUMBIA AND VENEZUELA

In Colombia and Venezuela, large coal deposits are found in numerous separate basins in intermontane valleys along the flanks and foothills of the Andean cordilleras and along their northern and eastern extensions. In Colombia, as shown in Figure 3, the coal measures range in age from Cretaceous to Tertiary, but the largest and most extensive deposits are of Late Cretaceous and Paleocene ages and are in the Guaduas Formation and its stratigraphic equivalents (Wokittel, 1960, p. 70). Few coalfields have been studied in detail, because of their large number, widespread distribution, and the small demand for coal. Rugged terrain and difficult access have hindered investigation of some.

The coal is in basins that range in area from less than 40 to several hundred square kilometres. According to Singewald (1950, p. 94–109), strata generally are folded and faulted; at places the structure is very complex. Dips of coal beds range from horizontal to vertical. In areas of complex structure, the coal commonly is pulverized and is colloquially referred to as "cisco." In a few areas, the coal has been metamorphosed by heat from igneous intrusions. Within individual coal districts, as many as 15 mineable coal beds, ranging in thickness from less than 1 m to as much as 8 m, have been found within stratigraphic intervals of 75 to 900 m.

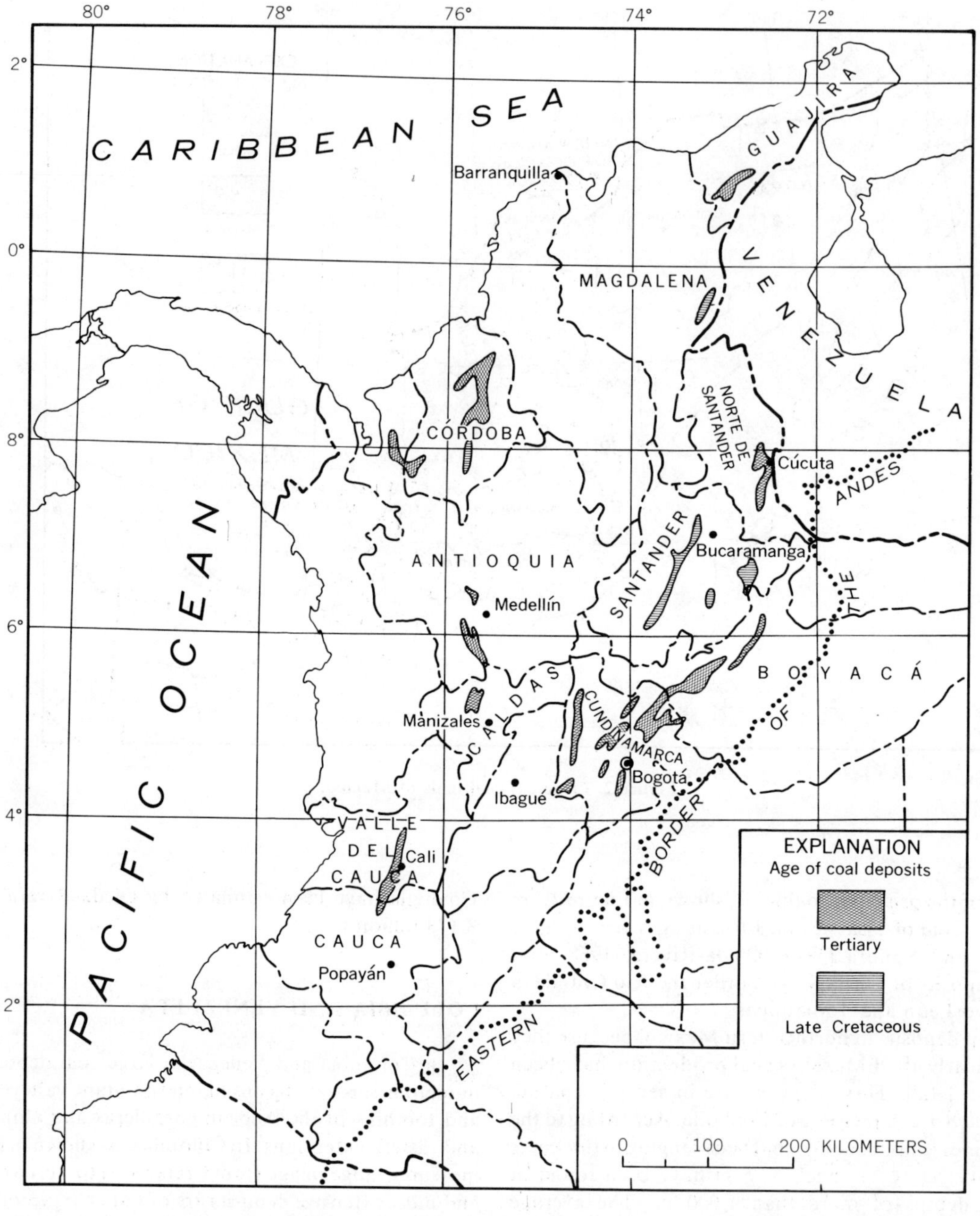

Figure 3. Principal coalfields of Colombia. Adapted from Singewald (1950, p. 8) and Wokittel (1960, p. 65).

Most of the coal is high-volatile bituminous and subbituminous, but coal of most other ranks is present. Ash and sulfur contents are generally low, but coal in Cauca and Valle del Cauca has a high ash content and a low to high sulfur content. Some of the bituminous coal produces a dense, high-ash, high-volatile coke that is used principally in blast furnaces and for domestic and commercial space heating.

Colombia contains the largest coal reserves of any Latin American country. Singewald (1950, p. 94) wrote, "The aggregate coal reserves in Colombia, to a maximum depth of a thousand meters, may very tentatively be regarded as 10 billion metric tons or more." Eighty percent of these reserves is in the departments of Cundinamarca and Boyacá. [Suescún-Gomez (this volume) placed reserves at 5 billion t, with a possible additional 5 billion t present. The total, 10 billion t, is the same as Singewald's estimate.]

The coal deposits in Venezuela, as described by Liddle (1928, p. 378) are in rocks of early Eocene and Miocene ages, as well as in rocks of Oligocene age (Marrero, 1964). The "early Eocene" as used in 1928 is probably equivalent to "Paleocene" of current usage. The oldest deposits, which are also the largest and most persistent, are in the lower part, commonly referred to by Liddle (1928, p. 181–208) as the "Third Coal horizon," of the Misoa-Trujillo

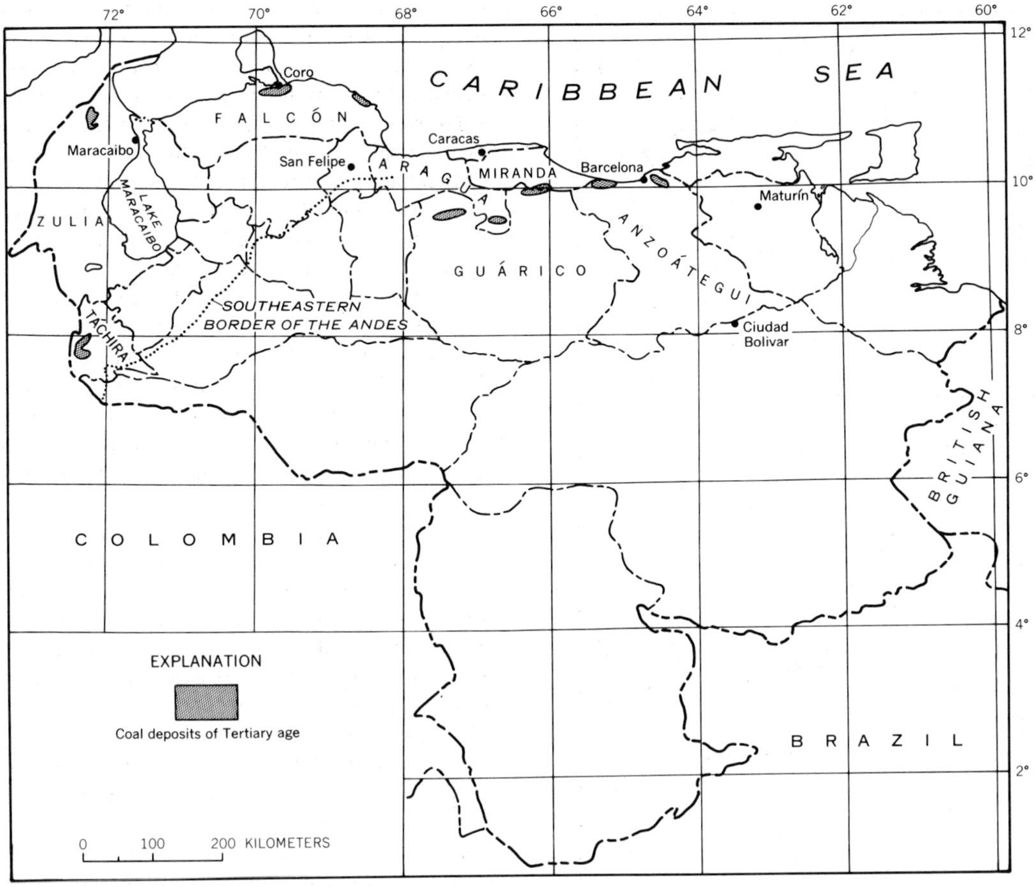

Figure 4. Principal coalfields of Venezuela.

Formation. The Third Coal horizon probably includes rocks that are stratigraphically equivalent, at least in part, to the Guaduas Formation of Colombia.

Within the principal coalfields of Venezuela (Fig. 4), the coal measures are as thick as 1,000 m (Liddle, 1928, p. 192). The number of coal beds within the sequence differs from place to place, but in general there are several. At one locality in Anzoátegui, Davey (1949a, p. 210; 1949b, p. 149) reported 20 coal beds more than 1 m thick, and at a locality in Zulia, he reported 15 coal beds ranging in thickness from 1.5 to 6.5 m. One and locally two coal beds, each 1 to 2 m thick, are being minded in Táchira. Coal beds in Falcón generally are no more than 1 m thick (Davey, 1950, p. 85).

The coal beds in all areas are folded and faulted. Dips as high as 70° to 85° are common in the Anzoátegui coalfields, but in Táchira, dips are locally as low as 8°. Coal beds in Zulia and Falcón are in shallow synclines and, in general, dip less than 35°.

The rank of Venezuelan coal ranges from lignite to semianthracite, but most is subbituminous and high-volatile bituminous. Ash content is generally less than 5 percent, and sulfur content rarely exceeds 1 percent. Some of the bituminous coal produces a weak coke that has limited use. A stronger, more satisfactory grade of coke is obtained by blending domestic coal with imported low-volatile bituminous coal.

Total reserves in six of the nine areas shown on Figure 4 are estimated by the U.S. Bureau of Mines (unpub. data) at a little more than 1 billion t of coal of all ranks.

PERU

The principal coalfields of Peru (Fig. 5) are found in the cordilleras of the Andes in the departments of La Libertad, Ancash, Lima, Pasco, and Junín. Numerous deposits are found along the Andes of southern Peru (Olive, 1959), but they are too small to have more than local value. In the department of Tumbes, in northern Peru, holes drilled during the late 1800s are reported to have penetrated several beds of lignite as thick as 4.5 m (Málaga Santolalla, 1921, p. 177–179).

Coal deposits of Peru range in age from Carboniferous to Tertiary, but all of the principal deposits are Cretaceous and are in the Goyllarisquizga Group or its stratigraphic equivalent (C. R. Petersen, 1968, oral commun.). The coal-bearing sequence ranges in thickness from 90 to 250 m. In most of the principal coalfields, the strata are intensely folded and faulted and are intruded by igneous rocks at many places.

In general, no more than three coal beds of mineable thickness exist at any one locality; as many as seven, however, ranging in thickness from 0.5 to 2 m, are reported by Brambilla (1957, p. 48) in a small valley in La Libertad. Beds 0.5 to 8 m thick are mined in northern Ancash. The coal beds commonly have moderate to steep dips and at places are overturned. Because of intense tectonism, the coal in some areas is pulverized.

The rank of the coal generally is bituminous, semianthracite, and anthracite. Bituminous coal is present in the departments of

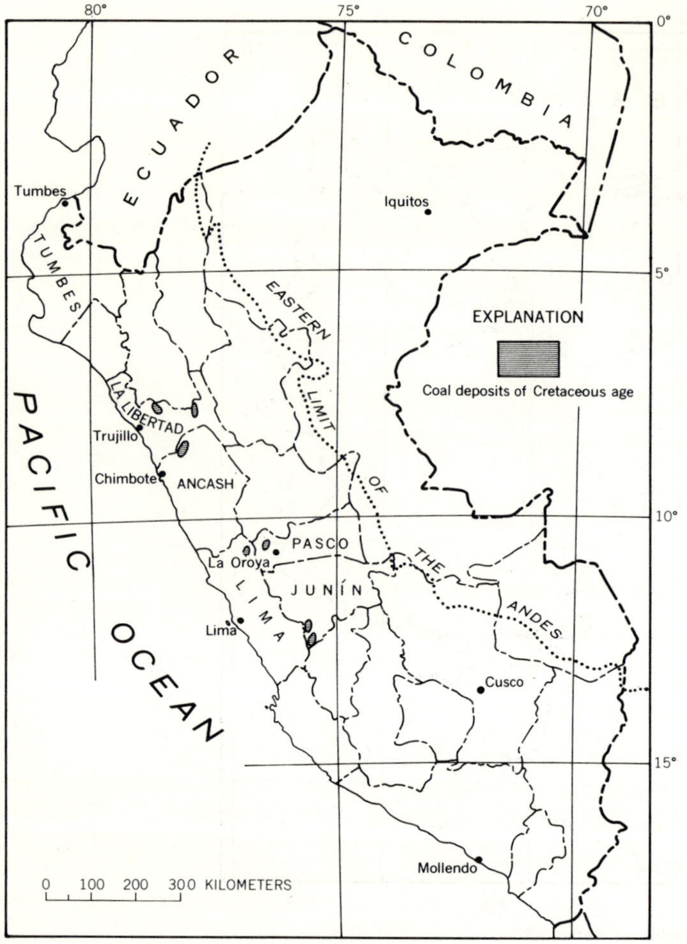

Figure 5. Principal coalfields of Peru.

Lima, Pasco, and Junín. Most of the semianthracite and anthracite deposits are in La Libertad and Ancash. The amount of ash in the bituminous coal ranges from moderate to great, and in anthracite and semianthracite, it is generally more than 10 percent. Sulfur content in all ranks is low. Much of the bituminous coal produces a coke that is used metallurgically.

Known coal reserves consist of 47 to 57 million t of anthracite in La Libertad and Ancash and 20 million t of bituminous coal in Lima, Junín, and Pasco. A very tentative estimate of total reserves is placed at 270 million t (C. R. Petersen, this volume) of coal of all ranks.

ARGENTINA AND CHILE

The principal coalfields of Argentina and Chile (Fig. 6) are between lat 36° S and 55° S. The deposits in Chile are near the Pacific coast; several in the provinces of Concepción and Arauco, near the town of Concepción, are entirely or almost entirely beneath the floor of the Pacific Ocean (Toenges and others, 1948, p. 5–30). In Argentina, the two principal coalfields are in the foothills of the Andes.

All the coal deposits are Tertiary in age and accumulated during Eocene and Oligocene time. Information on stratigraphic thickness is not available for all coalfields, but the sequence is known to be as thick as 90 m in the Concepción and Arauco coalfields (Juan Reccius, 1968, written commun.) and in the Pico Quemado coalfield (Borrello, 1956, p. 236). The sequence is as thick as 465 m in the Río Turbio coalfield (Borrello, 1956, p. 288). In some parts of the Magallanes coalfield, the upper part of the sequence has been eroded away, and glacial drift 2 to 12 m thick has been deposited above some of the coal beds (Toenges and others, 1948, p. 47).

The coal-bearing strata in the Concepción and Arauco coalfields of Chile and in the Río Turbio coalfield of Argentina dip monoclinally eastward and westward, respectively, at low to moderate angles. The coal measures in the Magallanes and Valdivia coalfields are folded into narrow synclines and anticlines. The Pico Quemado deposit is in a slightly asymmetric synclinal fold (Borrello, 1956, p. 233–234). Faults displace the coal-bearing strata in all areas and are numerous in the Concepción, Arauco, and Pico Quemado deposits.

Most deposits contain one to four coal beds that are more than 1 m thick. At least 10 coal beds having average thicknesses of 1 to 4 m and local thicknesses of as much as 8 m are reported (U.S. Bureau of Mines, unpub. data) in the Magallanes deposits. In coal mines of Chile and Argentina, no more than three coal beds, and generally only one or two, are exploited. Average thickness of beds in most mines is 1 m or more.

The rank of Chilean coal ranges from high-volatile bituminous A in the vicinity of Concepción to lignite and subbituminous in Magallanes. Coal in both Argentine deposits is ranked as high-volatile bituminous.

Concepción and Arauco coal contains generally less than 10 percent ash, whereas coal from mines in other parts of Chile contains 10 to 20 percent ash. Run-of-mine samples indicate that coal beds at Río Turbio and Pico Quemado contain from 15 to 35 percent ash, which is reduced considerably by washing. Sulfur content in Chilean coal is generally less than 2 percent and is 1 percent or less in Argentine coal. The coal is noncoking, but a suitable metallurgical-grade coke is obtained by blending Concepción and Río Turbio coal with imported high-grade, low-volatile bituminous coking coal.

Coal reserves in Chile were estimated by Wenzel (1965) at a little less than 4 billion t, of which 425 million t are measured and the remainder is indicated and inferred. [Flores-Williams (this volume) recalculated the total to be 5.9 billion t, of which 558 million t are measured, or proved.] Most of the reserves are in the Magallanes coalfield. About 5 percent, consisting mostly of high-volatile bituminous A coal, is in the Concepción and Arauco coalfields.

Total coal reserves in Argentina are estimated by Bergmann (this volume) at 484 million t. All but 5 million t are in the Río Turbio coalfield. Borrello (1956, p. 337) estimated Pico Quemado reserves at a little more than 2.5 million t. The remaining 2.5 million t occur in numerous, widely scattered, small deposits that range in age from Paleozoic to Cenozoic.

BRAZIL

The principal coalfields of Brazil (Fig. 7) are in the Brazilian Plateau, a deeply eroded tableland in the southern part of the country. According to Machado (this volume), the deposits are of

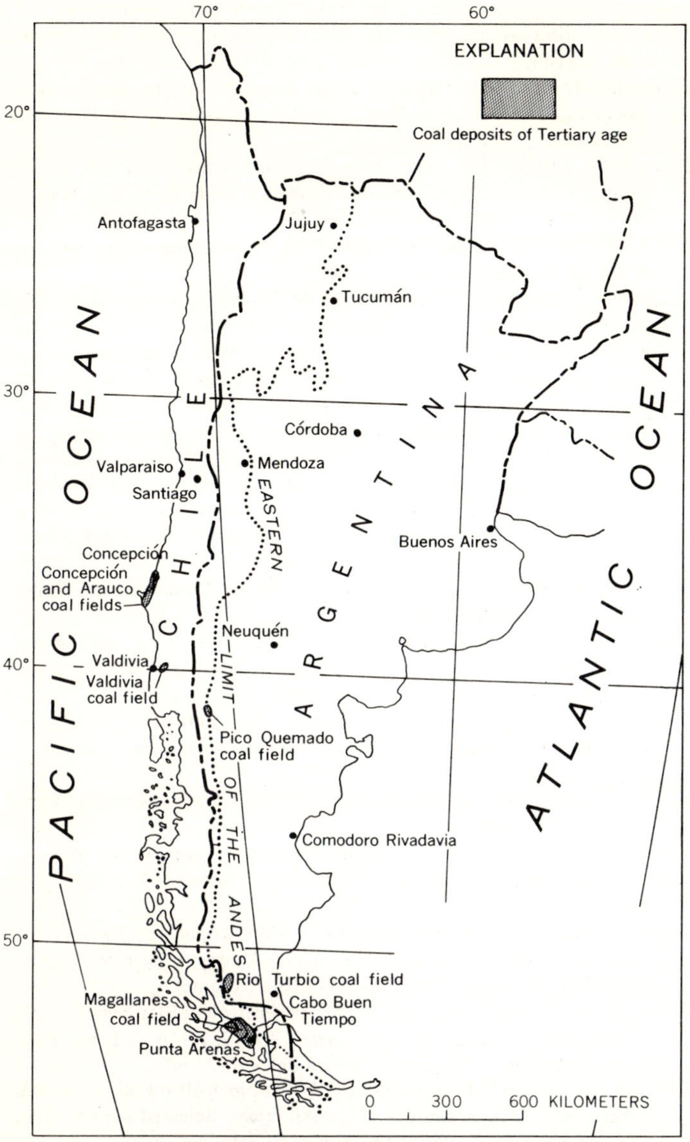

Figure 6. Principal coalfields of Argentina and Chile.

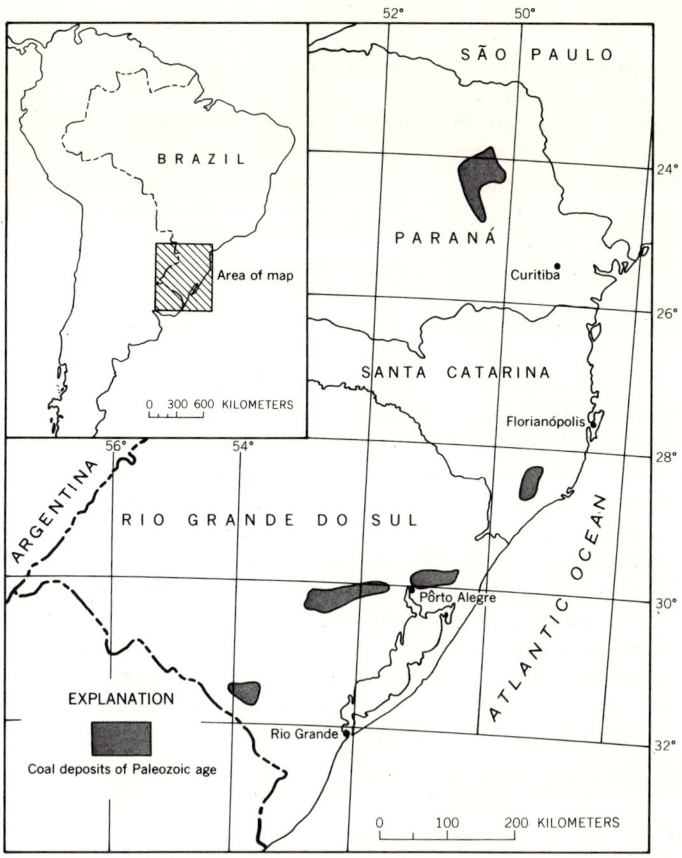

Figure 7. Principal coalfields of Brazil.

Early Permian and Pennsylvanian age and are in the Tubarõ Group. The coal measures accumulated in small isolated basins in areas peripheral to the Brazilian shield in the states of Paraná, Santa Catarina, and Rio Grande do Sul. According to Abreu (1962, p. 60), the stratigraphic position of the coal-bearing sequence differs from one basin to another. In Santa Catarina the sequence is as much as 70 m thick and contains four coal beds, or zones, that range in thickness from a few centimetres to 6 m (Abreu, 1962, p. 64, Fig. 11). Because of numerous partings, none of the thicker beds contains more than 2.25 m of usable coal. In Paraná the sequence contains two or three coal beds, each less than 1 m thick (Abreu, 1962, p. 65). Beds in Rio Grande do Sul are as thick as 3 m near the centers of basins but thin to 2 m or less near the basin edges.

The coal beds, in general, are flat-lying or gently dipping. At places they are broken by faults, some of which have large displacements. Coal beds in many areas have been intruded and metamorphosed by igneous rocks.

The rank of the coal ranges from lignite to anthracite; most is low- to medium-volatile bituminous. Small areas of semianthracite and anthracite coal border igneous intrusions.

Because of partings and impurities, the ash and sulfur contents of run-of-mine coal are generally large but are reduced considerably by means of beneficiation. Beneficiated coal from some areas is suitable for producing metallurgical-grade coke.

Most of the coal basins of Brazil have not been studied in detail, and data are not adequate to provide a reliable estimate of coal. Measured reserves in the states of Paraná, Rio Grande do Sul, and Santa Catarina do not exceed 1.885 billion t (Machado, this volume).

PRODUCTION

The leading coal-producing country in Latin America is Brazil, which produced annually between 5.2 and 5.9 million t during 1970 through 1973. Annual production in Mexico is about 4.25 million t, and in Colombia it is about 3.0 million t, but only about 60 percent of the coal is merchantable. Chile produces more than 1.4 million t annually, and Argentina, 625,000 t. Peru and Venezuela each produce less than 160,000 t annually.

IMPORTS AND EXPORTS

Brazil imported 1.8 million t of coal and coke during 1966, and Argentina imported 657,000 t during 1967. Chile, with an average annual import of 153,000 t, is the only other Latin American country that has been importing more than 100,000 t of coal annually. The amount of coal exported during the past decade from the coal-producing Latin American countries has been very small in comparison with world trade.

GOVERNMENT POLICIES

Government policy in most Latin American coal-producing countries generally is favorable to exploitation of coal resources by private enterprise except in Argentina. A law enacted in Argentina during 1958 declared hydrocarbon minerals to be the exclusive property of the State. In Mexico, concessions to mine coal are available, according to a 1963 law, only to Mexican citizens or to corporations that have 51 percent or more Mexican ownership. A law passed in Chile during 1965 authorizes the formation of joint mining companies involving private and government entities. These joint companies receive certain tax concessions and some exemptions that are not allowed other mining companies. Finally, the mining laws in Colombia presumably will be modernized (Suescún-Gomez, this volume).

CONCLUSIONS

Much remains to be learned about the thickness and distribution of coal-bearing sequences and about the characteristics of coal beds in Latin American countries. To meet future demands for energy and to provide a basis for wise investment and sound development, detailed information on the thickness, extent, attitude, and quality of coal beds, and on the thickness of overburden will be needed. To obtain such information will require areal mapping of coalfields at a scale of 1:50,000 or less, subsurface testing in areas of poor exposure, and analyses of coal from many localities. Only with such information can reliable estimates of recoverable coal reserves, according to rank, be made.

ACKNOWLEDGMENTS

I acknowledge with thanks the contributions and suggestions made by Paul Averitt, Ralph L. Miller, and Frank E. Noe.

REFERENCES CITED

Abreu, S. F., 1962, Recursos minerais do Brasil, *in* Combustíveis fósseis e minérios metálicos, Vol. II: Rio de Janeiro, Brazil Inst. Nac. Tecnologia, 696 p.

Bergmann, F.A.J., 1976, Coal resources of the Argentine Republic, *in* Kottlowski, F. E., and others, eds., Coal resources of the Americas: Selected papers: Geol. Soc. America Spec. Paper 179 (this volume).

Borrello, A. V., 1956, Recursos minerales de la República Argentina, III, Combustibles sólidos minerales: Buenos Aires, Inst. Nac. Inv. Ciencias Nat. Revista, v. 5, 665 p.

Brambilla, D. E., 1957, Geología de la zona antracífera de Huayday-Lucma, La Libertad: Soc. Geol. Perú Bol., v. 32 (Cong. Nac. Geología, 1st, Anales, pt. 2), p. 35–50.

Davey, J. C., 1949a, The Venezuelan Andes and the coastal and interior ranges. Part 3.—The eastern coast range: [London], Mining Mag. v. 80, no. 3, p. 137–144; no. 4, p. 204–213.

——1949b, The Maracaibo basin: [London] Mining Mag., v. 81, no. 1, p. 9–16; no. 2, p. 84–88; no. 3, p. 148–152.

——1950, Venezuela; the Falcón-Lara area: [London] Mining Mag., v. 82 no. 1, p. 9–20, no. 2, p. 80–88; no. 3, p. 150–152.

Evans, John, 1861, Geological report [coals of the Chiriqui district, Panama]: U.S. 36th Cong., 2d sess. House Executive Doc. 41, p. 45–55.

Flores-Williams, H., 1976, Chilean, Argentine, and Bolivian coals, *in* Kottlowski, F. E., and others, eds., Coal resources of the Americas: Selected papers: Geol. Soc. America Spec. Paper 179 (this volume).

González Reyna, Jenaro, 1956, Riqueza minera y yacimientos minerales de México (3rd ed.): Internat. Geol. Cong., México [D.F.] 20th, 497 p.

Jackson, C. T., 1861 [On coal from the Gulf of Chiriqui, Panama]: Boston Soc. Nat. History Proc., v. 7, p. 423, 428.

Liddle, R. A., 1928, The geology of Venezuela and Trinidad: Fort Worth, Texas, J. P. MacGowan, 552 p.

Machado, E. R., 1976, Coal in Brazil, *in* Kottlowski, F. E., and others, eds., Coal resources of the Americas: Selected papers: Geol. Soc. America Spec. Paper 179 (this volume).

Málaga Santolalla, Fermín, 1921, El carbón en el Perú: Perú, Cuerpo de Ingenieros de Minas y Aguas, Cong. Nac. Industria Minera Anales, v. 4, pt. 2, p. 48–332.

Marrero, Levi, 1964, Venezuela y sus recursos: Caracas, Cultural Venezolana, S.A., 700 p.

North American Geologic Map Committee, 1965, Geologic map of North America: Washington, D.C., U.S. Geol. Survey, 2 sheets, scale 1:5,000,000.

Ojeda-Rivera, J., 1976, Main coal regions of Mexico, *in* Kottlowski, F. E., and others, eds., Coal resources of the Americas: Selected papers: Geol. Soc. America Spec. Paper 179 (this volume).

Olive, W. W., 1959, Los recursos de carbón y petróleo: Plan regional para el desarollo del sur del Perú, Informes, Vol. 2, PS/A/6: Lima, Perú, Serv. Cooperativo Interam. del Plan del Sur, 24 p.

Petersen, R. C., 1976, Coal resources of Peru, *in* Kottlowski, F. E., and others, eds., Coal resources of the Americas: Selected papers: Geol. Soc. America Spec. Paper 179 (this volume).

Singewald, Q. D., 1950, Mineral resources of Colombia (other than petroleum): U.S. Geol. Survey Bull. 964-B, p. 53–204.

Suescún-Gómez, D., 1976, Coal deposits in Colombia, *in* Kottlowski, F. E., and others, eds., Coal resources of the Americas: Selected papers: Geol. Soc. America Spec. Paper 179 (this volume).

Toenges, A. L., and others, 1948, Coals of Chile: U.S. Bur. Mines Bull. 474, 106 p.

Wenzel, Osvaldo, 1965, Carbón, *in* Geografía económica de Chile: Santiago, Corporación de Fomento de la Producción (CORFO), chap. 20, p. 637–658.

Wokittel, Roberto, 1960, Recursos minerales de Colombia: Colombia Serv. Geol. Nac., Compilación Estud. Geol. Ofic. Colombia, v. 10, 393 p.

MANUSCRIPT RECEIVED BY THE SOCIETY DECEMBER 29, 1975
MANUSCRIPT ACCEPTED JANUARY 21, 1976

Printed in U.S.A.

Geological Society of America
Special Paper 179

Coal resources in Central America

Otto H. Bohnenberger
Gabriel Dengo
Instituto Centroamericano de Investigación y Tecnología Industrial (ICAITI), Guatemala City, Guatemala

ABSTRACT

A review of the available information on carbonaceous sedimentary rocks in the republics of Guatemala, El Salvador, Honduras, Nicaragua, Costa Rica, and Panama reveals the presence of coal and lignite beds, but no commercial production has been developed in any of these countries. The oldest coal beds, with thicknesses up to 2.5 m, crop out in central Honduras in at least six localities. They are interbeds in the argillaceous Triassic-Jurassic(?) El Plan Formation. This unit is strongly folded and faulted and in some places is intruded by dioritic plutons. The outcrops—in Yoro Department, Honduras—appear to be less deformed.

In west-central Guatemala, near the border with Chiapas State, Mexico, coal beds are reported that possibly are correlatable with the El Plan Formation. Thin lignite beds in Cretaceous carbonate rocks in Guatemala do not seem to be important.

In Panama, Costa Rica, and Guatemala, many different lignite beds, some more than 1 m thick, are reported in argillaceous and sandy Miocene strata. In Panama, these lignite beds occur in the Gatún Formation, whereas in the Izabal region of Guatemala, they are associated with beds of Miocene to Pliocene age. In most places, these Tertiary rocks are folded and faulted. They were evaluated in Izabal by a private mining company in 1968. Associated with sedimentary rocks of volcanic origin and probably of Pliocene to Pleistocene age, many thin lignite beds of local extent are found in intermontane basins of the volcanic provinces of the Central American countries. Only a few localities could attain importance for mining for local consumption.

INTRODUCTION

Our paper presents the available knowledge of carbonaceous sedimentary rocks of Central America, including the republics of Guatemala, Honduras, El Salvador, Nicaragua, Costa Rica, and Panama. Not a single deposit has been worked commercially. Our paper is based on a brief review of written information and not on field reconnaissance or on systematic sampling. The available information exhibits a wide range of reliability, ranging from old reports that give only localities to short, but more detailed, recent geologic reports. The information on carbonaceous sedimentary rocks is scarce and is scattered in a few printed papers, in mimeographed pamphlets of local distribution, or in internal office reports that are not readily available. In some countries, government agencies have devoted attention to carbonaceous beds, whereas in other countries almost nothing has been done.

The information available reveals that emphasis has been placed on proximate analysis of coal samples by some government agencies, although the sampling techniques themselves are never stated. Descriptions of the locations of the coal or lignite outcrops generally are imprecise. Almost nothing is presented in the available reports concerning related geologic factors such as thickness of overburden, extent of the deposit, type and rank of the coal, average thickness of the coal beds, structure of the coal-bearing beds, water-gas content, and amount of coal dust. As a demonstration of the paucity of information, all that was available is presented in Tables 1 through 9 for coals and lignites of each age group in Central America.

AGES OF CENTRAL AMERICAN CARBONACEOUS BEDS

The oldest sedimentary rocks in Central America—as dated by fossils—are in central Guatemala. These compose a sequence of a dark-colored shale, siltstone, and sandstone of Late Pennsylvanian to Early Permian age. No coal beds have been found in this sequence. In Honduras, lithologically similar, but younger, clastic sedimentary rocks occur, which have been dated as Late Triassic to Early Jurassic by means of plant remains (Newberry, 1888; Knowlton, 1918). This sequence, known as the El Plan Formation (Carpenter, 1954), contains coal beds in different localities. There is a possibility that older metasedimentary rocks in eastern Guatemala and central Honduras, where Paleozoic graphitic schists are known, contain some sheared or faulted coal beds, but there is little hope of finding a continuous bed in these metamorphosed rocks. In reconnaissance studies of the Honduras coal beds in the El Pataste and El Chile localities (Table 1), Hugh (1960) referred to them as Paleozoic coal on the basis of the discovery of some fossils found at El Pataste and identified by unknown individuals. However, regional correlation of the reported lithology of the coal sites suggests that these sedimentary rocks belong to the Triassic-

TABLE 1. MESOZOIC COAL DEPOSITS OF HONDURAS

Department	Name of prospect	Moisture (%)	Volatile matter (%)	Fixed carbon (%)	Ash (%)	Sulfur (%)	Calorific value
Yoro	Carbon Creek, tributary of Jacagua, east of Chilistagua; five beds each 0.5 m thick in a shale-sandstone sequence (Swanholm, 1967, p. 86–88)	32.1	28.5	24.3	15.1	2.5	5,970 Btu/lb
		21.1	27.6	31.2	20.1	3.0	7,200 Btu/lb
		23.8	29.9	33.4	12.9	3.1	7,940 Btu/lb
		30.2	25.7	20.2	23.9	2.2	5,330 Btu/lb
		26.3	17.3	15.4	41.0	1.8	..
Yoro	Unidentified	20.3	30.2	31.4	18.1	2.0	8,375 Btu/lb
Olancho	Río Frío (Laguna Seca), tributary of Rio Guayape, south of Catacamas: three beds 10 to 40 m apart in a slate sequence. Bed 1—1.80 m thick, which 1.27 m is accumulative coal, the rest being interbedded shale. Bed 2—2.7 m thick, of which 1.37 m is accumulative coal, the rest being interbedded shale. Bed 3—0.90 m thick, dirty coal (Elvir, 1959)	4.2	9.2	39.3	47.3	0.4	..
El Paraiso	El Pataste, 3 km north of El Pataste and 22 km north of Danlí: beds 2 to 5 m thick (Hugh, 1960, 1961a, 1961b)	1.60	15.81	30.58	52.01	1.24	4,401 cal/g
		0.75	20.87	18.62	59.76	1.03	2,725 cal/g
		13.00	43.53	3.78	62.69	0.51	400 cal/g
		0.75	18.83	13.96	67.21	1.19	2,224 cal/g
		1.20	18.47	19.77	61.76	3.76	3,763 cal/g
		1.68	17.56	49.50	31.26	0.36	10,443 Btu/lb
		1.50	18.20	55.92	24.38	0.49	11,523 Btu/lb
		13.50	17.80	23.60	45.10	0.50	4,675 Btu/lb
		15.60	18.90	25.20	40.30	0.40	5,140 Btu/lb
		13.30	18.40	25.60	42.70	0.40	4,970 Btu/lb
Francisco Morazán	El Chile, on southeast slope of El Chile Mountain, near La Meseta village, north of Moroceli: six very scattered beds as much as 2.5 m thick (Hugh, 1960, 1961a, 1961b)	13.10	12.70	32.00	42.20	0.30	..
		7.30	6.50	39.10	47.10	0.50	..
		15.90	18.20	32.90	33.00	0.30	5,770 Btu/lb
		12.05	2.29	44.68	40.98	0.15	6,373 Btu/lb
		12.59	2.77	43.15	41.49	0.15	6,674 Btu/lb
		4.40	9.50	32.00	54.00	0.60	4,940 Btu/lb
Unidentified	C. M. Cubas Turcios	3.60	5.10	81.90	9.40	0.70	12,510 Btu/lb
Francisco Morazán	Río Grande de Agua Caliente, Los Hozcones area, Municipalidad El Porvenir: one lignite bed 0.15 to 0.60 m thick, interbedded in gently dipping shale-siltstone sequence, overlain unconformably by glauconitic and silty sandstone (Elvir, 1967)	1.40	31.70	28.10	28.90	8.30	6,129 Btu/lb
Francisco Morazán	Río Siale–La Esperanza, 10 km northeast of El Porvenir: two beds 0.10 and 0.30 m thick, 15 m apart, in sandstone sequence (Swanholm, 1967, p. 86–88)				No analysis		
Francisco	Cerro Potrerillos, 9 km east of El Porvenir: coal occurrence reported but not checked (Swanholm, 1967, p. 86–88)				No analysis		

TABLE 2. MESOZOIC COAL BEDS OF GUATEMALA

Department	Name of prospect	Moisture (%)	Volatile matter (%)	Fixed carbon (%)	Ash (%)
Possibly of Jurassic age					
San Marcos	No name given	4.95	12.02	82.20	0.82
	No name given; near San José Ojetenán	0.56	56.30	26.01	18.50
	do.	..	34.31	41.21	24.48
			(sulfur 3.04%)		
Lignites alleged to be associated with Cretaceous sedimentary rocks					
Huehuetenango	Los Encuentros San Juan Ixcoy		No analysis		
Quiché	No name given; near San Juan Cotzal		No analysis		
Alta Verapaz	Carolina del Norte, Cobán	16.82	38.83	35.10	9.25
	Carolina del Sur, Cobán	13.35	36.20	22.15	28.30
	Wilhelm, Cahabón		No analysis		
	El Porvenir, Cobán		No analysis		
	El Chiu, Cobán		No analysis		
	Remería, Cobán	15.45	45.50	28.75	10.30
	Estrella del Sur, Cobán		No analysis		
Alta Verapaz	Reynería, San Miguel; San Miguel Tucurú		No analysis		
	Tucurú; San Miguel Tucurú		No analysis		

Note: The information of quantitative analysis of Guatemalan lignites and coal was provided by Julio Beltranena, Dirección General de Minería e Hidrocarburos, Guatemala City.

Jurassic(?) El Plan Formation. If the assignment to the El Plan Formation is correct, the oldest coal beds in Central America are in Honduras, and are of early to middle Mesozoic age.

An important group of lignite beds occurs in upper Tertiary clastic sedimentary rocks of marginal marine to continental depositional environments in northeastern Guatemala, Costa Rica, and Panama. These rocks range in age from late Oligocene to Miocene or possibly Pliocene.

TABLE 3. CENOZOIC LIGNITE BEDS OF PANAMA

Province	Name of prospect	Additional information
Bocas del Toro	No name nor precise location given; on islands of Laguna de Chiriquí	Lignite beds in Gatún Formation (Miocene)
Bocas del Toro	No name nor precise location reported	Do.
Veraguas	Near La Mesa and Santiago; precise location unknown	Do.
Herrera	Near Parita; no precise location reported	Lignite beds in Oligocene or Miocene sediments
Los Santos	Near Bombacho; no precise location reported	Do.
Colón	Río Indio; no precise location reported	Do.

Many so-called lignite beds and carbonaceous ashes are in volcanogenic Pliocene or even Pleistocene sediments in local basins in all of Central America. These carbonaceous beds have been grouped together as very young lignites.

Mesozoic Coal and Lignite Beds

Triassic-Jurassic(?). As stated in preceding paragraphs, Honduras has the best-known outcrops of lower to middle Mesozoic coal beds, interbedded in shale, siltstone, and sandstone sequences of the El Plan Formation. Some of these coal beds seem to be of good quality and attain thicknesses of more than 1 m. On the other hand, the rocks of the El Plan Formation have been subjected to large and repeated stresses and thus are strongly folded and faulted in most areas. In addition, in the El Pataste area, a dioritic pluton intrudes the sedimentary sequence. Because of these unfavorable conditions, Hugh (1961b) concluded that total Mesozoic (Paleozoic according to him) coal resources do not exceed 200,000 metric tons. Recent reconnaissance studies farther north, in Yoro Department (Table 1), showed that El Plan–type sedimentary rocks occur there and are less strongly deformed. This opens up possibilities of exploring for coal to the north, away from the main tectonic deformational axis of central Honduras.

In the department of San Marcos in west-central Guatemala, near the border with Chiapas State, Mexico, coal beds are reported that possibly are of similar age or slightly younger than the

TABLE 4. CENOZOIC LIGNITE BEDS OF COSTA RICA

Province	Name of prospect	Moisture (%)	Volatile matter (%)	Fixed carbon (%)	Ash (%)	Sulfur (%)	Calorific value (Btu/lb)
Alajuela	Mid-course of Río San Carlos	No additional information					
Alajuela	Region of El Venado, on north flank of Cordillera de Guanacaste, about 20 km northeast of Arenal Lake: Marcos Castro, no. 5	17.73	25.28	20.10	36.89	4.16	5,950
	Río La Muerte, no. 7	9.58	13.53	14.35	62.54	1.33	4,015
	Quebrada de Trevino, no. 9	24.33	23.56	20.39	31.72	2.14	6,740
	Danto, no. 2	22.92	22.72	12.89	41.46	2.69	4,578
	Danto, no. 10	26.38	21.93	11.47	40.22	1.68	4,458
	Raquel, no. 1	21.05	26.29	21.12	31.54	5.12	6,532
	Bed no. 13	19.66	15.64	12.63	52.07	2.26	3,606
	Bed no. 4	20.75	16.52	13.38	49.35	1.08	4,230
	Rafael Mora, no. 11	28.22	36.82	14.07	20.89	1.22	7,257
	Marcos Vargas, no. 2	28.07	26.40	19.98	25.55	1.25	6,729
	Quebrada de Burío, no. 10	18.07	18.16	5.36	58.41	1.81	..
Alajuela	Burío, no. 8	20.82	15.87	2.08	61.23	0.22	..
	El Túnel, no. 1, bed no. 6	9.77	9.42	6.99	73.83	0.93	..
Limón	Zent, 32 km southeast of Puerto Limón: Zent 1	31.55	39.39	19.73	9.33	8.93	8,730
	Zent 2	30.25	47.22	16.64	5.89	0.61	10,257
	Zent 3	23.87	48.29	24.13	3.71	0.42	10.700
	Zent 4	24.43	45.34	26.20	4.03	1.09	10,350
	Zent 5	23.85	42.48	20.09	15.58	3.90	8,830
	Zent 6	24.15	42.48	25.28	8.09	2.36	10,889
	Zent 7	28.38	42.47	21.13	8.02	4.74	10.215
Limón	Catarata de Puerto Viejo, 12 km west of Puerto Viejo	Lenticular lignite beds, about 0.60 m thick, in siltstone-sandstone sequence of Gatún Formation (Miocene)					
San José	El Tablazo Hills, 2 km south of El Higuito	Lenticular lignite beds as much as 0.80 m thick, in siliceous sandstone; poor quality (Umaña, 1964)					

Note: Laboratory analysis by S. Bravo Pérez, 1965, San José, Costa Rica.

TABLE 5. CENOZOIC LIGNITE BEDS OF GUATEMALA

Name of prospect	Moisture (%)	Volatile matter (%)	Fixed carbon (%)	Ash (%)	Sulfur (%)	Calorific value (Btu/lb)
Alaska, Izabal	18.72	36.97	34.40	10.51	..	6,000
Crique Chino, 200 m northwest of Crique Chino Creek	35.52	29.53	23.63	11.32	5.16	10,087
Chino Creek	14.85	30.80	25.00	29.35	4.49	5,673
Chino Creek	15.20	37.02	28.23	19.55	7.90	6,662
Chino Creek	26.38	34.68	25.28	13.66	4.22	..
El Carmen, Livingston	11.69	39.05	41.10	8.16	..	..
Río Salado, 3 km southwest of Bahía de Amatique; 0.35-m-thick bed, horizontal			No analysis			
La Casualidad, Livingston	1.23	35.22	50.99	12.56	..	7,000
La Reina, Izabal	3.28	32.00	62.32	2.40	..	8,000
La Romana, Río La Romana, 200 m south of Las Pavas River, 3 km west of Bahía de Santo Tomás	9.43	33.45	47.69	9.53	..	..
Santo Tomás	8.98	40.54	38.73	11.75	8.52	10,300
Providencia, Livingston			No analysis			
Río Carbonera, 10 to 12 km south of Lake Golfete; 11 lignite beds totaling 8 m(?)	42.17	28.37	17.00	12.46	3.30	..
Carbonera, Río Carbonera, 10 to 12 km south of Lake Golfete; 11 lignite beds totaling 8 m(?)	21.10	25.90	24.36	28.64	2.63	..
Río Frío, about 2 km south of Lake Golfete; 2 beds 0.43 and 0.28 m thick, respectively	22.74	30.66	29.02	17.58	..	6,354
Río Lámpara, 6 km southwest of Livingston, Izabal	12.09 13.77	37.20 38.00	33.79 33.79	14.44 14.44	.. 4.97	.. 10,165
Río Lámpara	25.73	27.93	17.34	29.00	5.30	5,308
Río Lámpara	22.33	18.99	50.76	7.90	..	..
San Carlos, 4 km west of Bahía de Amatique; two beds 0.30 m and 0.35 m thick, respectively	5.30	33.15	25.20	36.27	4.97	7,397
Río Santa Elena, 4 km southwest of Bahía de Amatique; two almost horizontal beds 0.30 and 0.35 m thick, respectively	15.11	34.08	40.20	10.61	..	..
Río Tameja near Río Dulce; three beds 0.10 m thick	6.35	39.62	33.53	20.60	..	10,043

Note: All in Izabal Province.

TABLE 6. VERY YOUNG LIGNITE BEDS OF NICARAGUA

Department	Prospect
Chinandega	Las Mercedes; claim by Rafael Espinoza and others
Chinandega	Providencia; claim by Rafael Espinoza and others
Chinandega	San Luis; claim by Rafael Espinoza and others
León	Margarita; claim by A. Sánchez Salinas
León	Providencia; claim by Roberto D. Amort and others
León	Soledad; claim by A. Sánchez Salinas
León	Paula; claim by A. Sánchez Salinas
Estelí	Region of Estelí; La Laguna, El Cacao, Chomotillo, and La Concordia
Madriz	Valley between Pueblo Nuevo and Somoto
Chontales	Río Mico; precise location unknown

Note: No additional information is available.

TABLE 7. VERY YOUNG LIGNITE BEDS OF HONDURAS

Department	Prospect	Additional information
Francisco Morazán	Sabana Grande near Rancho Maldonado, south of Sabana Grande; one bed as much as 1.80 m thick, dipping 30° and more in siliceous shales, unconformably overlain by volcanic rocks (Hugh, 1960)	Moisture, 11.0% volatile matter, 25.1%; fixed carbon, 43.3%; ash, 20.6%; sulfur, 1.8%; and calorific value, 8,680 Btu/lb
Valle	Curaren, about 10 km east of Aramecina; shaly lignite (Hugh, 1960)	None
Yoro	Las Flores, no precise location reported only briefly mentioned (Hugh, 1961b)	None
Ocotepeque	In the Sensenti Valley, precise location unknown (Hugh, 1960)	None

Honduras coal, or at least suggest a correlation with Jurassic rocks (Table 2). In San Marcos, these sedimentary rocks are covered by thick Tertiary and Quaternary volcanic rocks, and little is known about stratigraphic relations in this area.

Cretaceous. On Figure 1, the Mesozoic age symbol also has been used to indicate those coal or lignite localities in Guatemala that are in Cretaceous sedimentary rocks (Table 2). The Cretaceous strata are predominantly geosynclinal carbonate rocks, and thus the claims that carbonaceous beds occur in the Cretaceous area are questionable. Regardless of their age, some of these lignite beds were evaluated for local use by a mining company about 30 years ago and were found to be noncommercial. They are included in the listed localities because some of them warrant a new appraisal.

TABLE 8. VERY YOUNG LIGNITE BEDS EL SALVADOR

Department	Prospect	Ash* (%)
Santa Ana	No name given; near mouth of Río Tahuilapa into Río Lempa	81.2
Cuscatlán	San Pedro Perulapán, Cantón Istagua	..
	Tenancingo, Cantón Ajuculco	6.4
	Suchitoto, Cantón Tenango	..
Cabañas	Ilobasco, El Hoyo	49.3
	de La Labor	73.8
	Río Los Frailes, Ilobasco	71.35
	(different beds)	10.9
		65.0
		65.3
		76.8
		51.3
		69.1
		82.3
		81.6
		86.2
		82.0
		66.8
		41.4
		52.3
		81.5
		69.1
		83.4
		87.0
		51.3
		85.8
	Río Las Cañas, Ilobasco	..
	Río Los Naranjos, San Isidro	..
	Valley of Río Lempa, north of Jutiopa	..
	Valley of Río Lempa, east of Nombre de Jesus	..
	Valley of Río Lempa, near San Juan Lempa	..
	Valley of Río Lempa, near Santa Lucia	..
San Miguel	Neuvo Edén de San Juan	..
	No name given; west of Carolina	83.5
	(different beds)	84.4
	Carolina, north of Río Torola	53.2
	(different beds)	53.0
Morazán	San Simón	..
	No name given; north of Yoloaiquín	75.4
La Unión	El Carmelo, south of Yucuaiquin	64.5
	(different beds)	58.0

*Leaders (..) indicate no data.

Cenozoic Lignite Beds

In Panama and Costa Rica, Tertiary time is well represented by a fairly complete succession of fossiliferous sedimentary rocks. Throughout this period, but mainly during Miocene time, local basins were formed with the proper conditions for the accumulation of carbonaceous beds (Tables 3, 4). All reported localities seem to be confined to the upper part of the Gatún Formation of Miocene age, where lignite and lignitic siltstone are interbedded with argillaceous and sandy sedimentary rocks. In most of the reported localities, the Gatún Formation is strongly deformed, with dips ranging from 10° to 40°. The lignite beds are lenticular. These factors have hindered commercial use of the lignite beds, some of which individually seem to be of good quality.

A similar although slightly younger sequence of argillaceous sedimentary rocks occurs in Guatemala in Izabal Department, near the Caribbean coast, and near the potential inland waterways of Río Dulce, Lake Golfete, and Lake Izabal (Table 5). Numerous thin lignite beds are reported to be present in the siltstone, sandstone, and mudstone of a unit to which no formal name has been given. The consensus is that these rocks range in age from late Miocene and Pliocene. The main problem, aside from the difficult access, is the intense local deformation of the lignite beds. Again, areas must be sought where folds are gentle and cutoffs by faults are not so abundant. A private mining company, after evaluating the lignite beds of the Izabal region in Guatemala, abandoned exploration activities.

Very Young Lignite Beds

During the late volcanic history of Central America, probably in Pliocene and Pleistocene time, local intermontane basins were formed by block faulting or by a combination of tectonic events and volcanic eruptions. In some of these basins, organic material accumulated along with volcanic ash or with siliceous precipitates. Thus, many occurrences of so-called coal beds are reported in the volcanic regions of Central America. These very young lignite beds, or lignitic ash beds, are thin, lenticular, and of only local extent. Some localities, however, warrant a systematic evaluation; in this category we include the localities of Sabana Grande (Francisco Morazán) and Las Flores (Yoro), both in Honduras. Most of the lignite claims of this group in Guatemala, El Salvador, and Nicaragua are probably of no commercial value. On the other hand, these lignite beds are the only ones in Central America not deformed by former orogenies, and they should be investigated systematically (Tables 6 through 9).

GENERAL COMMENTS ON REPORTED ANALYSES

Our tabulations of coal or lignite proximate analyses (Tables 1 through 9) were drawn from many different sources, some of which are not indicated specifically, and include private and governmental laboratories in Central America, as well as private and government research facilities abroad. The results reflect the sample as received by the laboratory. No one knows how the sampling was performed, how much the sample was dried, and how much time elapsed between the sampling and the analysis. Miller (1961) wrote

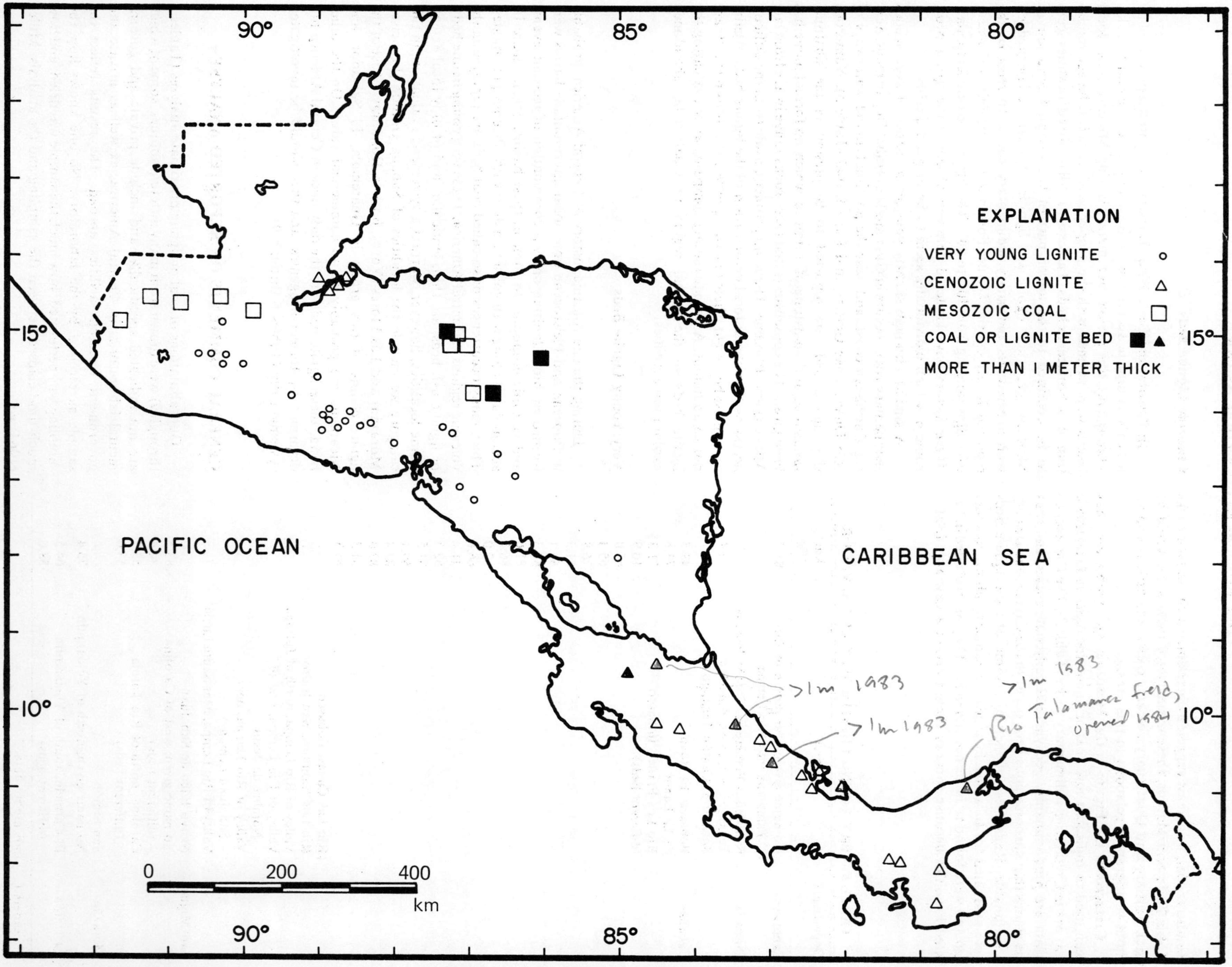

Figure 1. Coal and lignite localities in Central America.

TABLE 9. VERY YOUNG LIGNITE BEDS OF GUATEMALA

Department	Prospect	Moisture (%)	Volatile matter (%)	Fixed carbon (%)	Ash (%)
Guatemala	El Tablón, San José Pinula		No analysis		
	Gracias a Dios, Palencia		No analysis		
	No name given	2.55	19.00	62.70	15.70
	Nueva Wellington, San Juan Sacatepéquez		No analysis		
	Tudela, Chinautla		No analysis		
	El Bellotal, San Raymundo		No analysis		
El Progreso	Sansare, Sansare		No analysis		
	No name given; Aguacatillo	0.23	0.04	90.12	
			(sulfur 0.50%)		
	No name given; San Antonio La Paz	20.80	64.84	12.06	2.30
	No name given; Río Los Plátanos		No analysis		
Sacatepéquez	No name given; Santo Domingo Xenacoj		No analysis		
Chimaltenango	No name given; Patzún		No analysis		
	No name given; San Martín Jilotepeque		No analysis		
	No name given; Rio Quisayá		No analysis		
Quezaltenango	No name given; Quezaltenango		No analysis		
Baja Verapaz	Vega Larga; Salamá		No analysis		
Jalapa	No name given; Jalapa	16.38	29.07	37.40	16.70
	Trinidad; Mataquescuintla	—22.50—		77.50	7.00(?)
	Sansoro; Mataquescuintla	9.20	30.34	0.15	60.28

In order to have absolute or comparative value, a coal sample must be taken in a rigidly prescribed, meticulous manner, and must be hermetically sealed to prevent loss of moisture between the time of sampling and the time of analysis. It is extremely unlikely that the analyzed samples were taken in this manner, and hence the original moisture content of the coal is probably not correct.

From this it can be concluded that a systematic campaign to evaluate the carbonaceous beds of Central America should be carried out by adequately trained geologists who should sample all coal beds in a standard manner and take the necessary precautions to furnish the laboratory with a representative sample at the time of the analysis. The results of such a program would be extremely valuable for a regional evaluation. Finally, all other geologic factors affecting a coal deposit should be investigated during the same exploratory program.

CONCLUSIONS

Coal or lignite beds occur in sedimentary rocks of early Mesozoic, late Tertiary, and very late Tertiary to Quaternary age. Not enough information has been gathered to prove or disprove the commercial value of any deposit. Regions of gentle deformation should be sought in the first two age groups; exploration for coal or lignite beds should be concentrated in these areas through a coordinated, systematic evaluation of the entire Central American region. A systematic inventory of the very young lignite beds should eliminate from consideration all but a few localities, which could be evaluated further.

ACKNOWLEDGMENTS

We are particularly indebted to Ing. Reniery Elvir, Honduras, Dr. Eduardo Rodezno, Nicaragua, and Ing. Enrique Malavassi, Costa Rica, for critically reading the manuscript. We likewise appreciate the preliminary editing of the original manuscript in 1968 by William Hunter. A. A. Meyerhoff kindly edited the final version.

SELECTED REFERENCES

Note: During the years 1965 to 1967, the AID published a series of atlases showing the physical resources of the Central American countries. These very cumbersome volumes (size about 24 in. by 30 in. by 4 in.) were prepared by the Resources Inventory Center (RIC) as a General Inventory of Physical Resources (GIPR) for the Agency for International Development (AID) Regional Office for Central America and Panama (ROCAP) and executed by the Corps of Engineers, U.S. Army, Bureau for Latin America, Rural Development Division (RDD). Sometime in the 1960s, the former International Cooperation Administration (ICA) changed its name to the Agency for International Development; therefore, the older publications still carry the former name.

Panama

Agency for International Development Resources Center, 1967, National inventory of physical resources, Central America and Panama: Washington, D.C., Bureau for Latin America ID/RDD and ROCAP, Agency for Internat. Devel., U.S. Army Corps Engineers, August, AID/RIC GIPR no. 7, 52 p.

Costa Rica

Agency for International Development Resources Inventory Center, 1965, Regional analysis of physical resources, Central America and Panama; Costa Rica: Washington, D.C., Bureau for Latin America, ID/RDD and ROCAP, Agency for Internat. Devel., U.S. Army Corps Engineers, August, AID/RIC GIPR no. 4, 83 p.

Bravo Perez, S., 1966, Estudio sobre las características y composición de los lignitos de Zent y Venado: Costa Rica Univ. Pubs., Facultad de Ciencias y Letras, Departamento de Química, Tesis de Grado, 75 p.

Umaña, J. E., 1964, Informe de progreso en la investigación y evaluación de yacimientos de lignito en Costa Rica: San José, Inst. Costarricense de Electricidad, unpub. rept., p. 2–4.

Umaña, J. E., Escalante, G., and Malavassi, E., 1964, Yacaimientos de lignito en la región de Venado: San José, Inst. Costarricense de Electricidad, Sección de Geología y Departamento de Geología, Minas y Petróleo, Ministerio de Industrias, August, mimeo. rept., 14 p., 3 pls.

Nicaragua

Anónimo, 1957, Resumen de los reconocimientos de las zonas mineralizadas y otros estudios: Nicaragua Servicio Geol. Nac. Bol., no. 1, p. 10.

Lykkeberg, E. R., 1954, Riqueza minera Nicaraüense, Resumen del Catastro Minero: Managua, D.N., Ministerio de Economía, unpub. rept. p. 8–14.

Honduras

Carpenter, R. H., 1954, Geology and ore deposits of the Rosario mining district and the San Juancito Mountains, Honduras, Central America: Geol. Soc. America Bull., v. 65, 23–28.

Crouse, L. F., 1956, Informe de los estudios preliminares de las potencialidades de carbón en la República de Honduras: Tegucigalpa, Internat. Coop. Adm., Am. Embassy, October, unpub. rept., 25 p.

Elvir, A., R., 1959, Informe geológico preliminar sobre los yacimientos de carbón de Río Frío, Jurisdicción de Catacamas, Olancho: Tegucigalpa, Banco Central de Honduras, March, unpub. rept., 8 p.

——1967, Exploración de carbón, area de los Horcones, Municipio El Porvenir, Dept. de Francisco Morazán: Tegucigalpa, Dir. General de Minas e Hidrocarburos, August, unpub. rept., 5 p.

Hugh, K. E., 1960, Estudio de las posibilidades le existencia de carbón en Honduras: Tegucigalpa, Banco Central de Honduras, August, unpub. rept., 3 p.

——1961a, Estudio geológico de los yacimientos de carbón de "El Pataste" y de la Montaña "El Chile": Tegucigalpa, Banco Central de Honduras, January, unpub. rept., 5 p.

——1961b, Resultado de la investigación de las posibilidades carboníferas en Honduras, hecha bajo el patrocinio del Banco Central de Honduras: Tegucigalpa, Bol. Mensual, Banco Central de Honduras, v. 11, no. 5, May, 2 p.

Knowlton, F. H., 1918, Relations between the Mesozoic floras of North and South America: Geol. Soc. America Bull., v. 29, p. 607–614.

Newberry, J. S., 1888, Rhaetic plants from Honduras: Am. Jour. Sci., ser. 3, v. 36, p. 342–351.

Swanholm, J., 1967, Honduras mineral exploration project, final rept.: Tegucigalpa, United Nations Tech. Assistance Operations, Dir. General de Minas e Hidrocarburos, Ministerio Recursos Naturales, December, 108 p.

El Salvador

Agency for International Development Resources Inventory Center, 1965, Regional analysis of physical resources, Central America and Panama; El Salvador: Washington, D.C., Bureau for Latin America ID/RDD and ROCAP, Agency for Internat. Devel., U.S. Army Corps Engineers, February, AID/RIC GIPR no. 3, 70 p.

Grebe, W. H., 1955, La minería en El Salvador, Centro América: San Salvador, El Salvador Anales Servicio Geol. Nac. Bol., no. 1, March, p. 33–36.

Guatemala

De Buen, R., 1950, Estudio sobre aprovechamiento de los carbones minerales de Guatemala y en especial de los del Departamento de Izabal: Guatemala City, Inst. Fomento y Producción, August, unpub. rept., 23 p.

Miller, R. L., 1961, The mineral position of Guatemala with special reference to industrial minerals: U.S. Geol. Survey Open-File Rept., Internat. Coop. Adm., April, 66 p.

Manuscript Received by the Society February 3, 1975
Manuscript Accepted March 17, 1975

Printed in U.S.A.

Geological Society of America
Special Paper 179

Main coal regions of Mexico

Jesús Ojeda-Rivera
Cerro Gordo 373, Col. Campestre Churubusco, México, 21 D.F., México

ABSTRACT

Most of the coal in Mexico is concentrated in three widely separated regions: Coahuila on the northeast, Sonora on the northwest, and Oaxaca on the south-southeast. In the Coahuila region, bituminous coking coals are present in Upper Cretaceous rocks. The Sonora region contains deposits of Upper Triassic anthracitic coal, altered by probable thermal metamorphism. The Oaxaca region contains anthracitic and some bituminous coking coal in Lower(?) to Middle Jurassic sequences.

Through 1973, only the coal deposits of the Coahuila region had been developed commercially. The coals of the Sonora and Oaxaca regions are undeveloped because of unfavorable geoeconomic conditions and lack of exploration. Through 1973, the Coahuila region had yielded about 110 million metric tons of coal, most of which was used to produce coke. Present production is 4,250,000 metric tons per year. The largest mines include one underground mine (Sabinas no. 2) and one openpit mine (La Florida), which produce 40,000 metric tons per month.

Only reconnaissance-geologic investigations, complemented locally by a few limited drilling projects, have been made in the Sonora and Oaxaca regions. Exploration has been more extensive in the Coahuila region. There are many totally unexplored areas with potential coal resources.

The estimated resources in the Sabinas area of the Coahuila region total 1,781 million metric tons classed as reserves in situ, and 1,180 million metric tons classed as potential but not commercially important at present. North of the Sabinas area, in the Río Escondido basin, the estimated resources total 120 million metric tons classed as measured.

In the Sonora region the estimated resources total 11 million metric tons classed as reserves in situ for the Santa Clara coalfield and 4 million metric tons classed as measured, 9 million metric tons classed as inferred, and 18 million metric tons classed as possible for the San Marcial district.

In the Mixtepec and Tezoatlán coal fields of the Oaxaca region, the estimated resources total 24 million metric tons classed as inferred and 26 million metric tons classed as possible.

GEOGRAPHIC FEATURES AND GEOMORPHOLOGY

Most of the coal in Mexico is concentrated in three widely separated regions: Coahuila on the northeast, Sonora on the northwest, and Oaxaca on the south-southeast (Figs. 1 through 4).

The coal deposits of the Coahuila region lie in a geomorphic province characterized by moderately high mountainous areas on the west that pass gradually eastward into isolated ranges, separated by broad alluvial plains and pediments. The climate is hot and dry. The winters are mild, although from time to time both summers and winters may be extremely hot or cold. Rainfall is sparse. Two permanent rivers cross the region: the Salado River on the south and the Sabinas River on the north. A desert type of vegetation in the lower areas grades gradually into shrubs and trees at higher elevations.

In the Sonora region, coal deposits are in a geomorphic province of parallel ranges and valleys, characterized by north-

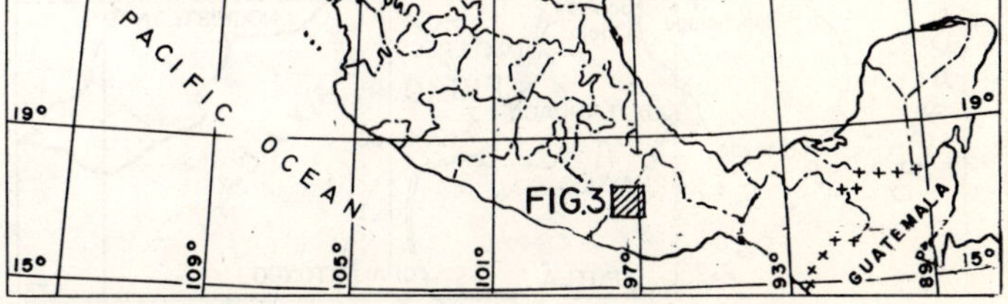

Figure 1. Index map of main coal-bearing regions of Mexico. Locations of Figures 2, 3, and 4 are shown.

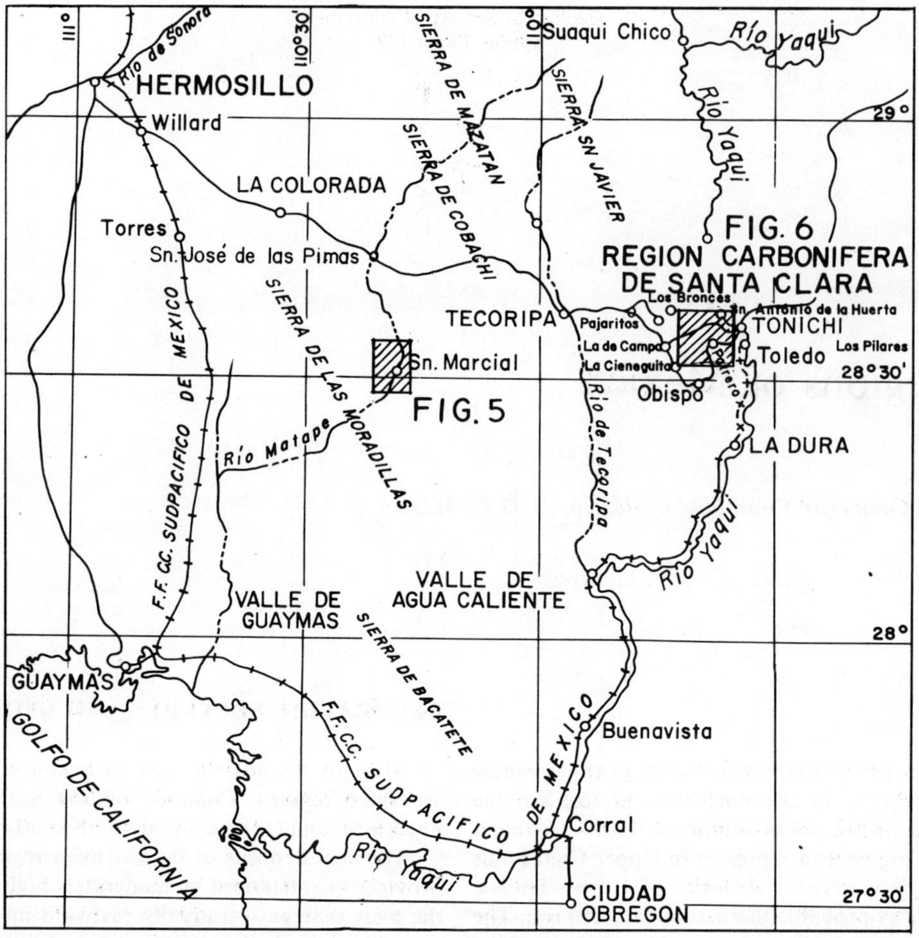

Figure 2. Index map of coal-bearing area of Sonora, Mexico. Locations of Figures 5 and 6 are shown.

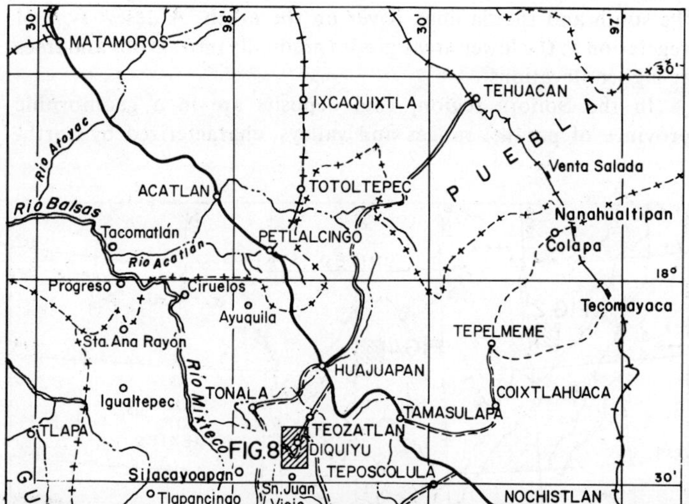

Figure 3. Index map of coal-bearing area of Oaxaca, Mexico. Locations of Figures 8 and 10 are shown.

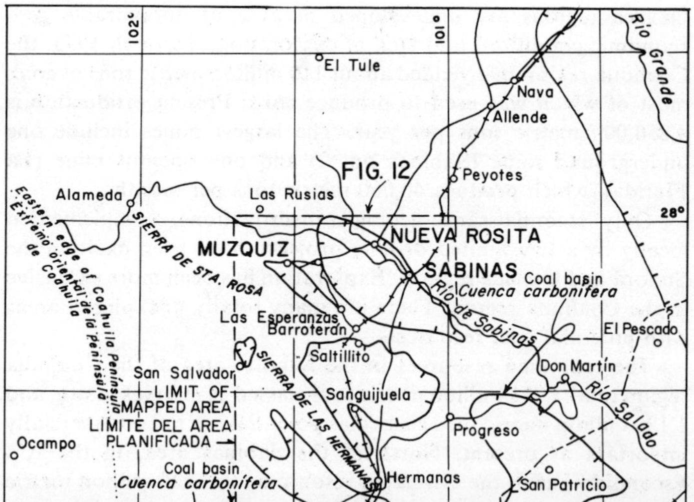

Figure 4. Index map of coal-bearing area of Coahuila, Mexico. Location of Figure 12 is shown.

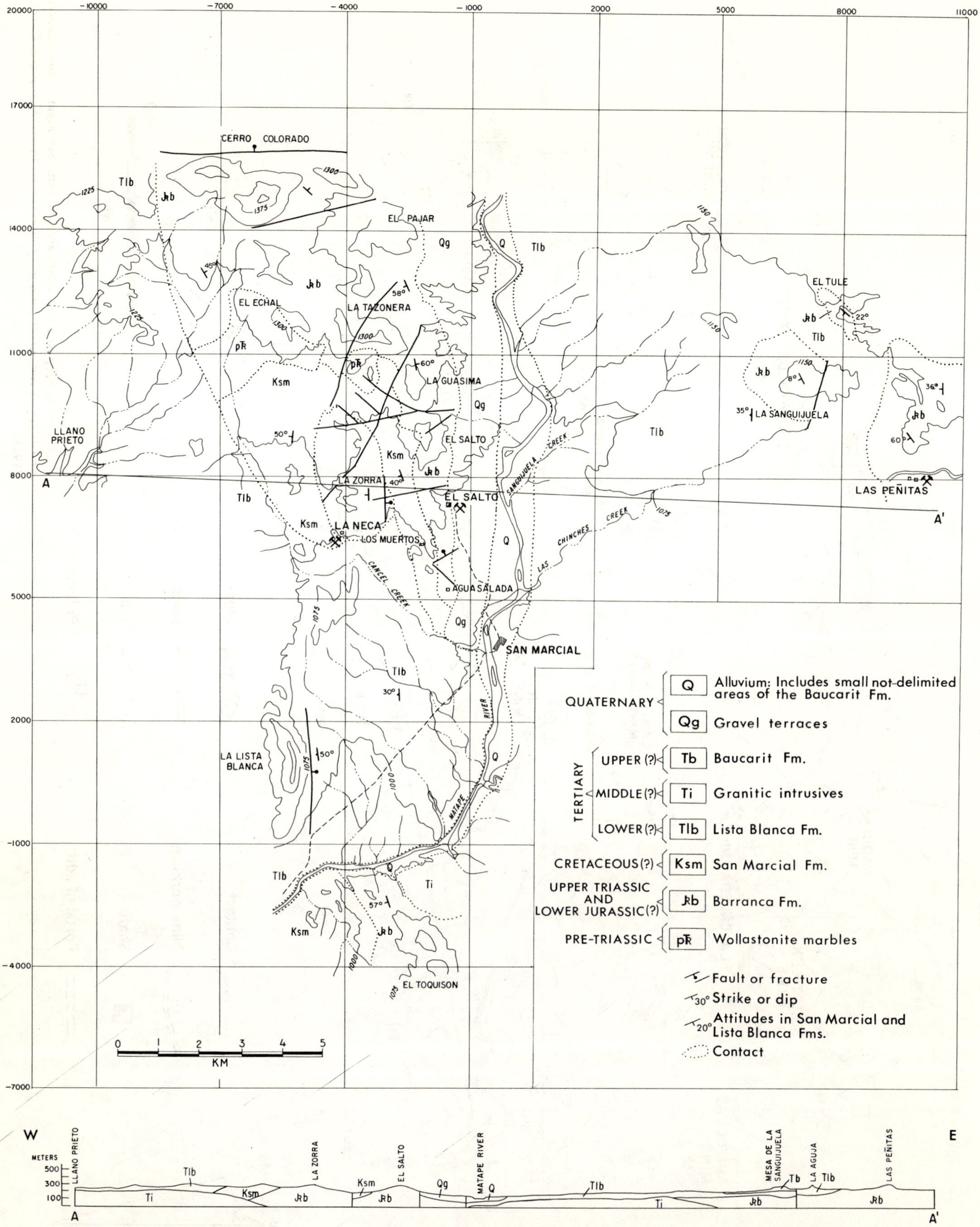

Figure 5. Map of San Marcial area, Sonora. See Figure 2 for location. From Pesquera and Carbonell (1960).

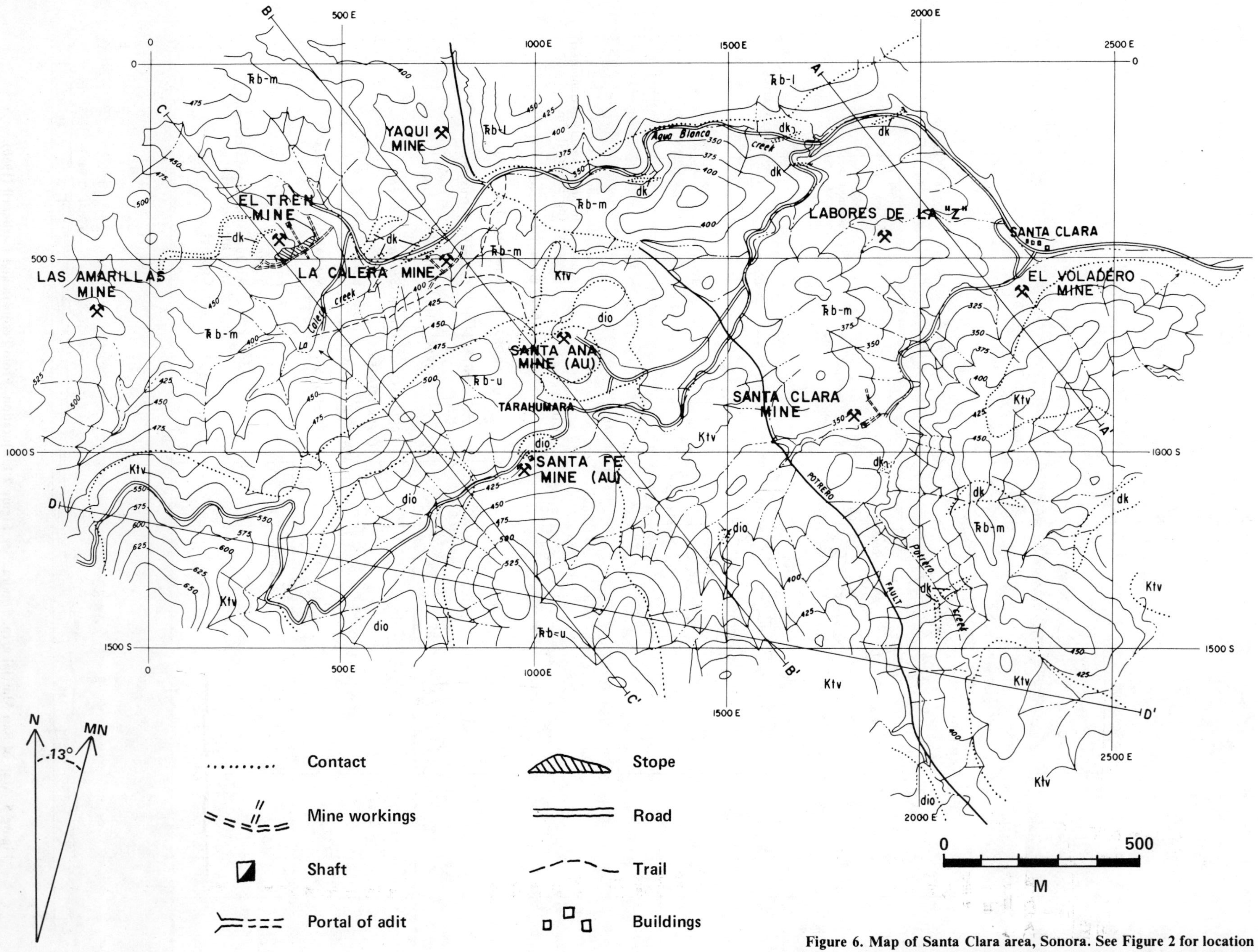

Figure 6. Map of Santa Clara area, Sonora. See Figure 2 for location and Figure 7 for explanation of geologic units. From Wilson and Rocha (1964).

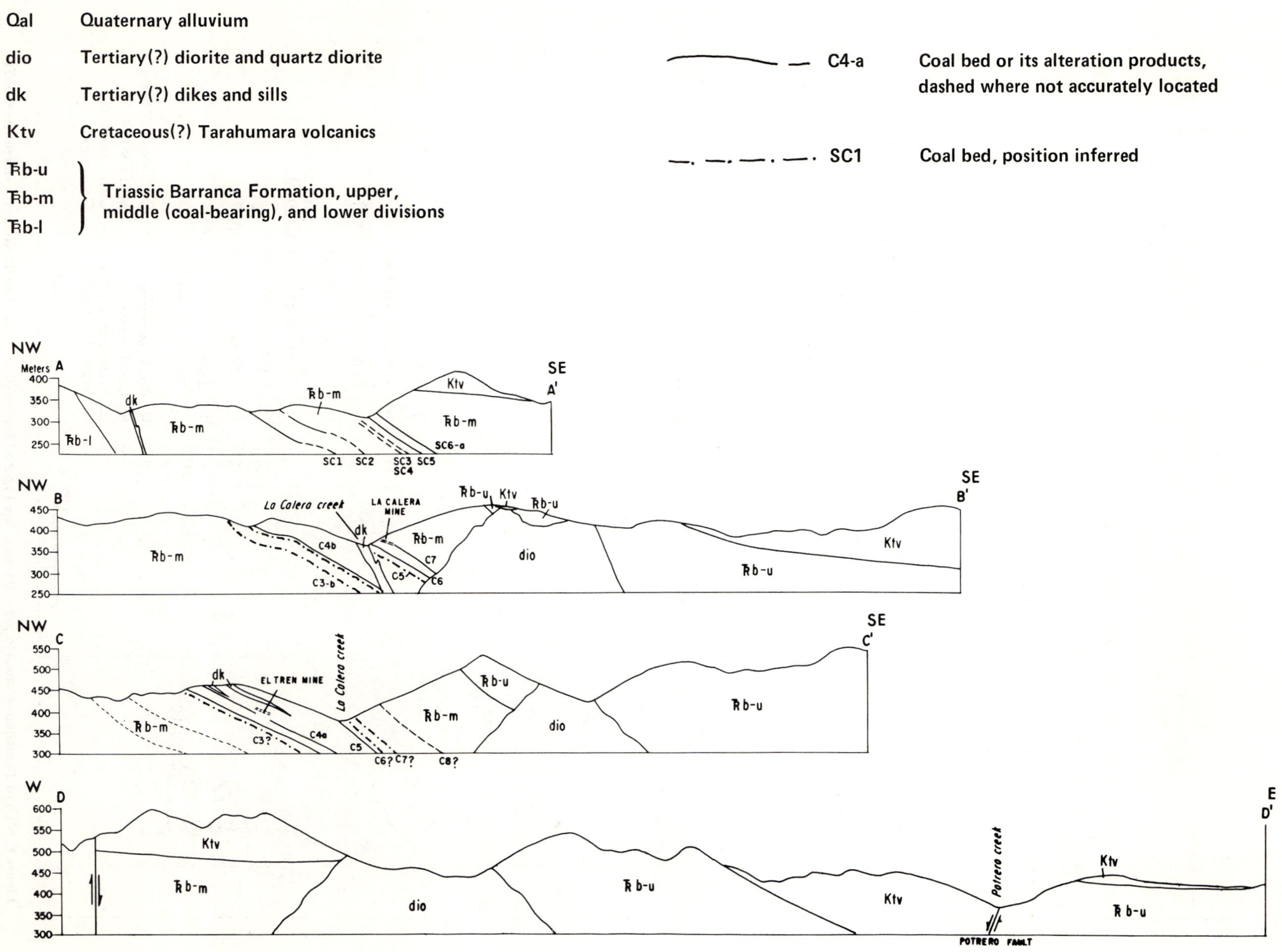

Figure 7. Cross section, Santa Clara area. See Figure 6 for location. From Wilson and Rocha (1964).

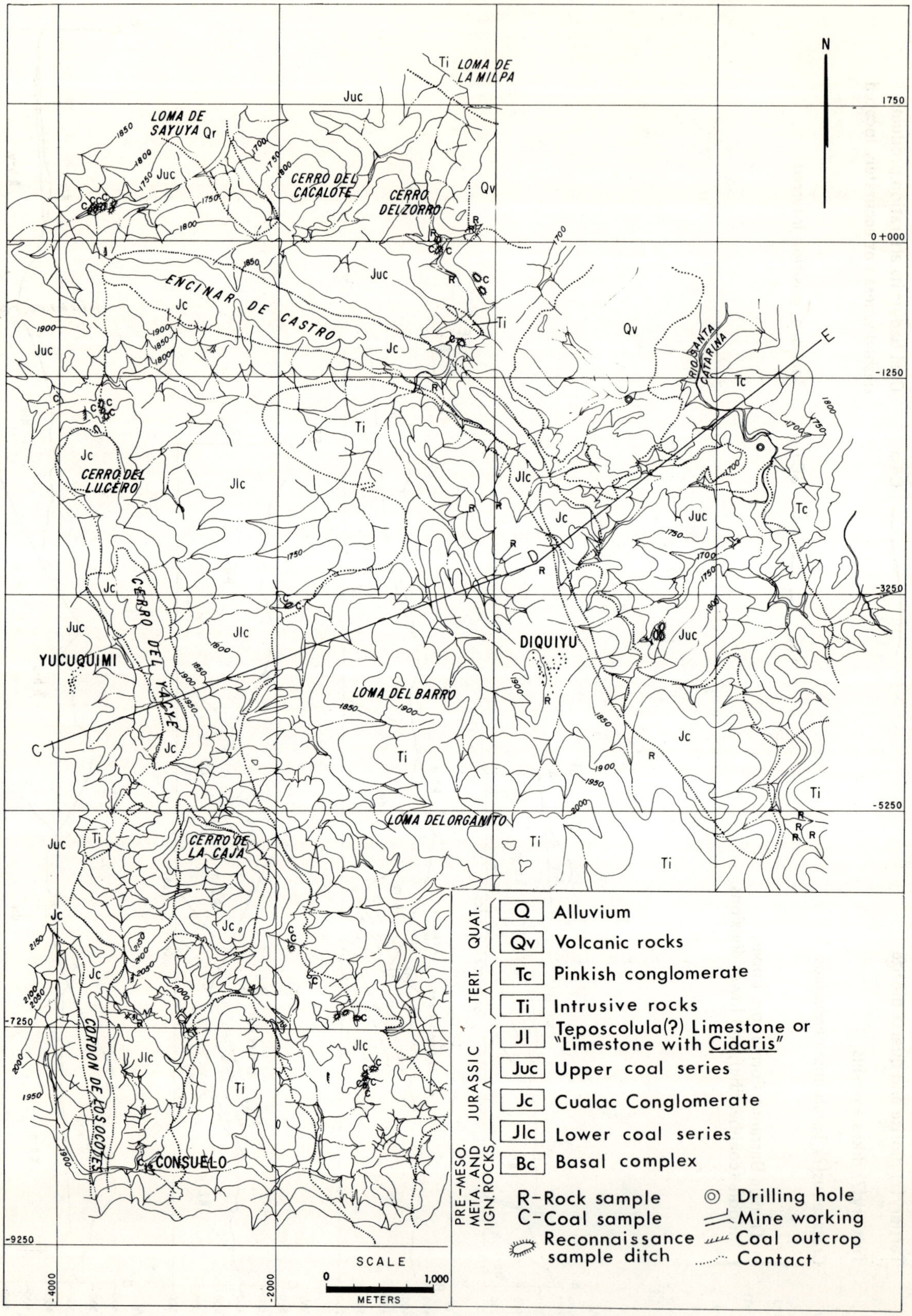

Figure 8. Map of Tezoatlán-Consuelo area, Oaxaca. See Figure 3 for location. From Cortés-Obregón and others (1957).

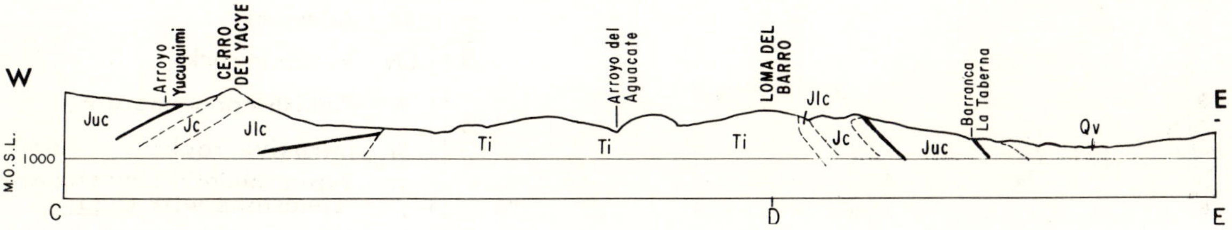

Figure 9. Cross section, Tezoatlán-Consuelo area, Oaxaca. See Figure 8 for location; explanation as in Figure 8.

trending, fairly long mountain ranges of moderate elevation, separated by narrow longitudinal valleys. The only permanent river that crosses the region is the Yaqui River, which passes 7 to 10 km east of the Santa Clara coal district. Another important but intermittent stream is the Matape River, which runs directly across the San Marcial coal district. The climate is hot and dry. The winters are mild, but occasionally the temperature falls below 0° C. The summers are hot, and temperatures up to 40° C are not uncommon. The rainy season generally begins in the latter part of June and usually ends in October. Most of the region is thickly covered with brush, cactus, low shrubs, and thorny deciduous trees of several varieties.

The coal deposits of the Oaxaca region are on the east-northeast margin of the Sierra Madre del Sur in a geomorphic province locally known as the Sierra de la Mixteca. This region is characterized by extremely mountainous and rugged topography. The climate is warm, and the winters are mild. The rainy season begins in May or June and ends in October. Two permanent rivers cross the region: the Mixtepec and Salado Rivers, which drain the Mixtepec and Tezoatlán coal districts, respectively. A subtropical type of vegetation at lower elevations changes abruptly into desert and forest types at intermediate and higher elevations.

The accessibility and economic development of these regions are quite different. Coahuila is an easily accessible, important, industrial mining and agricultural center, whereas Sonora and Oaxaca are poorly accessible, sparsely inhabited areas with only a few small towns and some dirt roads which, in some places, may be impassable during the rainy season.

GEOLOGIC SKETCH OF COAL-BEARING REGIONS

Sonora Region

In the Sonora region, coal-bearing rocks in the middle part of the Upper Triassic Barranca Formation crop out in two small areas of a few square kilometers. These are the San Marcial area, about 90 km southeast of Hermosillo, and the Santa Clara area, about 60 km east of San Marcial (Figs. 5 through 7).

The oldest rocks crop out in the Sierra de Covachi, northeast of La Casita, and consist mainly of limestone beds in which Ordovician and Permian fossils have been found. Most of the ranges, including those near the Santa Clara and San Marcial areas, are underlain principally by rocks of the Barranca Formation of Late Triassic and perhaps Early Jurassic age. The Barranca Formation consists of a thick sequence, partly marine and partly nonmarine, of quartzitic sandstone, conglomerate, shale, and intercalated coal beds.

The Barranca Formation is unconformably overlain by two volcanic complexes, one of which is probably of Cretaceous age and the other probably of Tertiary age. The volcanic sequences include local sills and dikes of probable Cretaceous age, and extensive stocklike bodies of probable Tertiary age.

In the Santa Clara area (Figs. 6, 7) the rocks of the Barranca Formation have a general northeastward strike and a southeastward homoclinal dip averaging 30°. Coal beds occur in two basins, the Calera and Santa Clara basins, separated by the Potrero fault. The Calera basin contains at least nine coal beds and the Santa Clara basin at least seven; each bed is at least 1 m thick in some places. In most areas, weathering has completely destroyed the coal to depths of 10 to 30 m.

In the San Marcial area (Fig. 5), the Barranca Formation and overlying rocks form isolated, north-trending, upfaulted blocks that dip steeply westward. Ten coal beds, four of them probably commercial, lie in an area of at least 2.5 sq km.

The coal has been highly metamorphosed to a rank between anthracite and meta-anthracite. It has a higher specific gravity, although it is more friable, than typical anthracite. The most troublesome impurity is graphite. Natural coke has developed in places near igneous intrusive bodies. The coal is characterized by a very low hydrogen content, fairly high moisture, and moderate to fairly high ash.

This coal has been used in some silver and copper smelters. Although not suitable for use in locomotives, it could be used in stationary boilers or domestic heating units through the development of special burning equipment.

Coal Reserves of Sonora

The coal reserves of Sonora are difficult to estimate because of the lack of exploration and the complicated geology. Reports by Wilson and Rocha (1964) and Camacho and others (1956) suggest that reserves in the Santa Clara area total 11,731,010 metric tons (hereafter, t), of which 252,558 are classed as measured and 6,756,101 are classed as recoverable. Pesquera and Carbonell (1960) estimated that reserves in the San Marcial area total 4 million t classed as measured, 9 million as inferred, and 18 million as possible.

Although these deposits may appear to be small, as well as difficult to explore and develop, they are the only known coal resources near the northern Pacific coast of Mexico. As such, they deserve more exploration and experimentation to make mining both feasible and profitable.

Oaxaca Region

Coal outcrops in Oaxaca are widely scattered across an area of more than 10,000 sq km. However, only two relatively small areas,

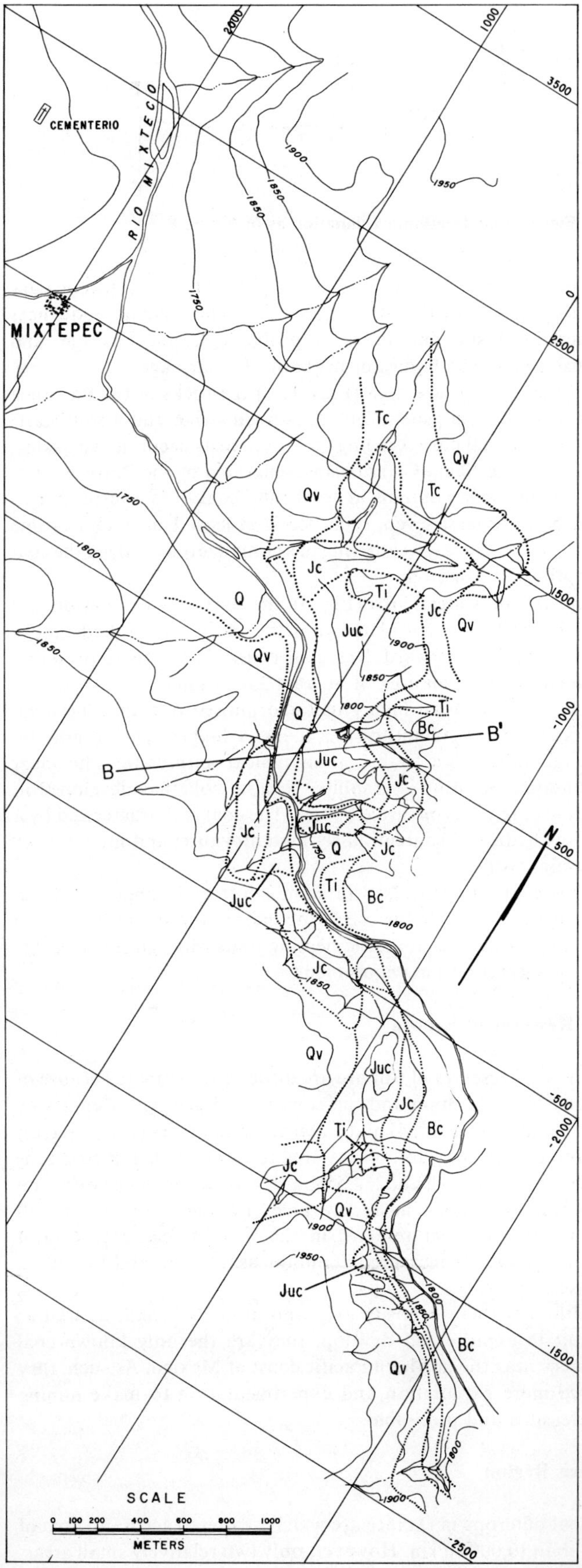

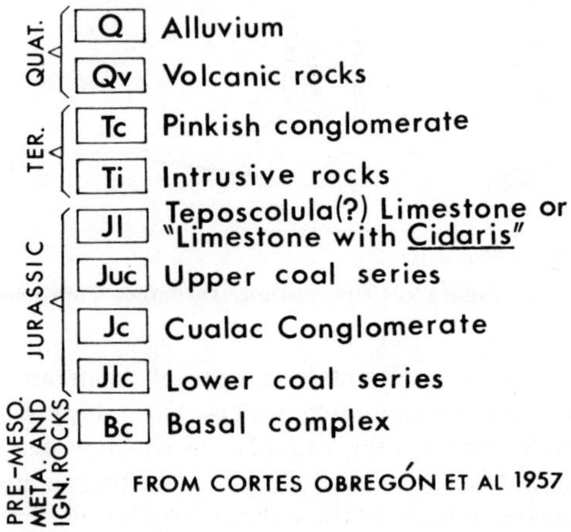

Figure 10. (left) Map of Mixtepec area, Oaxaca. See Figure 3 for location; explanation above.

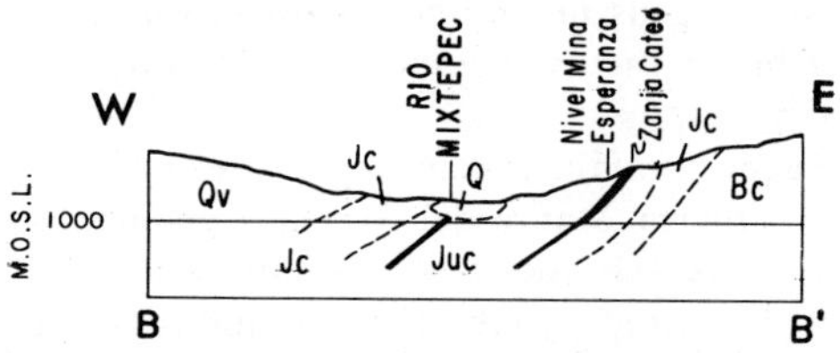

Figure 11. Cross section, Mixtepec area, Oaxaca, See Figure 10 for location and explanation.

where the coal-bearing rocks are best exposed and most accessible, have been explored in some detail. These are the Tezoatlán area (100 sq km), which is about 40 km west of Huajuapan (Figs. 8, 9), and the Mixtepec area (30 sq km), which is about 25 km south of Tezoatlán (Figs. 10, 11). The oldest rocks cropping out in this region probably are of pre-Permian age. They are the "basal complex," which consists of metamorphic rocks, mostly metasedimentary, showing different grades of metamorphism and including intrusive bodies of different ages and composition.

Resting unconformably on the basal complex are 3,000 to 4,000 m of sedimentary rocks, in part marine and in part nonmarine, ranging in age from Early(?) Jurassic to Holocene and in composition from quartzitic sandstone and conglomerate, siltstone, and shale with intercalated coal beds to fossiliferous limestone, dense limestone, conglomerate, water-laid pyroclastic rocks, and alluvium.

The main coal-bearing units are the Rosario, Zorrillo, and Simón Formations of Early(?) to Middle Jurassic age, which consist of intercalated sandstone, siltstone, and shale, including coal beds ranging in thickness from a few centimeters to 3 m.

At some places the Cualac Conglomerate, a quartz conglomerate 30 to 80 m thick, rests unconformably on the Rosario Formation and conformably underlies the Zorrillo Formation.

In the Tezoatlán area, a nucleus of post-Mesozoic intrusions is present in the "basal complex." These igneous rocks initially were intruded along the axis of a probable N60° W-trending anticline. The Jurassic beds lie on the periphery of the uplifted and eroded basal complex and have been intensely folded and faulted.

The Mixtepec area (Figs. 10, 11) also shows the effects of post-

Mesozoic intrusions, but to a much lesser degree than the Tezoatlán area. Although the Jurassic beds have been intensely folded and faulted, a northeast strike characterizes both the structures and the sedimentary beds.

Because of the structural complexity in both areas, coal-bed attitudes and thicknesses are quite variable. In general, dips range from 30° to 50°; only very locally are they less than 20°. Coal-bed thicknesses are so variable that they rarely are constant across distances greater than 50 m. Coal outcrops are scarce; the coal beds are difficult to correlate; and in places the coals have been destroyed completely by weathering to depths of 10 to 30 m.

The coal of the Rosario Formation is of low-volatile bituminous and semianthracite rank with no coking properties. The coal of the Zorrillo Formation in the Mixtepec area is of medium-volatile bituminous rank with a certain degree of coking properties not shown by coal of the Zorrillo Formation of the Tezoatlán area, probably because of the proximity of igneous intrusions.

A common property of coals from both areas is a high ash content, which is difficult to remove by washing. However, some successful laboratory-scale experiments have been made to improve their quality.

Coal Reserves of Oaxaca

Coal resources reported by Cortés-Obregón and others (1957) total 126,378,624 t, of which 18,530,688 t are of coal showing a certain degree of coking properties. Of these totals, 13,115,456 t is classified as measured, 52,718,048 as inferred, and 60,545,120 as possible. However, the last exploration carried out in 1973–1974 by the Consejo de Recursos Naturales No Renovables in the Mixtepec and Tezoatlán areas reports a total of only 24 million t classified as inferred and 26 million as possible.

Coahuila Region[1]

The main coal deposits of the Coahuila region are in the Sabinas area (Fig. 12) where the Sabinas, Las Esperanzas, and Saltillito basins have for a long time been economically the most important. The rest of the known basins of the Sabinas area are the Lampacitos, San Patricio, Las Adjuntas, San Salvador, and Monclova basins, which extend toward the south-southeast over more than 9,000 sq km. Under the present (1974) conditions of development, none of these basins, with the possible exception of a few areas owned by private companies and a small coal lens found at Las Adjuntas, has proved to be of commercial importance.

The exposed sedimentary rocks, about 5,000 to 6,000 m thick, range in age from Late Jurassic to Holocene and in composition from limestone and shale of marine origin to red beds and conglomerate; they include coal and gypsum deposits.

The most economically important units of the sedimentary sequence are in the Upper Cretaceous section, which includes mainly equivalents of the upper Taylor (San Miguel Formation) and the Navarro (Olmos, Escondido, and Múzquiz Formations). The lithology of these units reflects alternate marine and nonmarine conditions of deposition. Coal is present in the Olmos Formation, which consists of nonmarine siltstone and sandstone 366 to 498 m thick; the coal beds are lenticular. Although the Múzquiz coals are interbedded in rocks as old as those in the middle part of the Olmos Formation, they are mainly equivalent to the Escondido Formation.

Resting unconformably on the Upper Cretaceous rocks are remnants of the Pliocene(?) Sabinas Conglomerate. In the western half of the area, conspicuous lava flows of the Pleistocene Esperanzas Lava overlie the Cretaceous strata in the Sabinas, Las Esperanzas, and Saltillito basins. Upper Pliocene to Pleistocene playa-lake sediments underlie the Holocene alluvial valley fill toward the centers of the basins and, in much of the Las Adjuntas and Sabinas basins, conceal the Cretaceous strata.

Tertiary intrusive bodies are known in nearby areas on the southwest and may be present at shallow depths in some of the mountain ranges on the northwest. However, within the limits of the coal basins, only a small quartz-diorite stocklike body, the Loma de Farías, crops out on the south in the junction zone between the Las Adjuntas and Monclova basins.

The geologic structure is rather simple and consists of several northwest-trending mountain-forming anticlines and adjacent basin-forming synclines, with only minor local complications. These structures constitute part of the Coahuila marginal folded province of northeastern Mexico.

The flat coastal plain in the northeast grades into the mountainous plateau on the southwest with increasingly large anticlines and synclines. Thus, northeast of the Sierra de Hermanas, the synclincal coal basins are shallow and elliptical, whereas those on the southwest are deep, long, and narrow.

Marine conditions prevailed through Cretaceous time until the end of Taylor deposition, which ended with the deposition of a white beach sand, on which a double coal bed was deposited in an area paleogeographically known as the "Coahuila coal basin." Marine and continental conditions alternated until the end of the Cretaceous Period, after which only continental conditions existed and erosion took place as folding and uplift progressed.

The double coal bed can be mined commercially where the rock parting is less than 10 to 20 cm thick and the two coal benches have a total thickness of at least 2 m. However, in most of the marginal areas of the basins, splitting or thinning out of the double bed is fairly common, and even in the best "pockets" it averages nearly 30 percent ash. The coal is of medium-volatile bituminous rank, and most of it can be used to make metallurgical coke.

Mining began in the San Felipe area of the Sabinas basin in 1884 and has been continuous since 1889. Thirteen companies have been operating in the Sabinas, Las Esperanzas, and Saltillito basins since 1956. La Florida, an openpit mine, and Sabinas no. 2, an underground mine—the largest mines—have a monthly production of 40,000 t. The present yearly production is about 4,250,000 t, and plans call for a production of 10 million t by 1980. To date, the region has produced about 110 million t, most of which has been used to make coke.

Coal Reserves of Coahuila

According to Robeck and others (1956), "original" resources in the Sabinas area, including the deeper or less accessible zones of the basins, are as much as 12,279,900,000 t. Of this total, 4,128,634,000 t are at the margins of the basins or at relatively shallow depths. The amount of recoverable coal, calculated from four categories of

[1] Figures on coal reserves in the Coahuila region were being revised by the author but were not available when this book went to press.

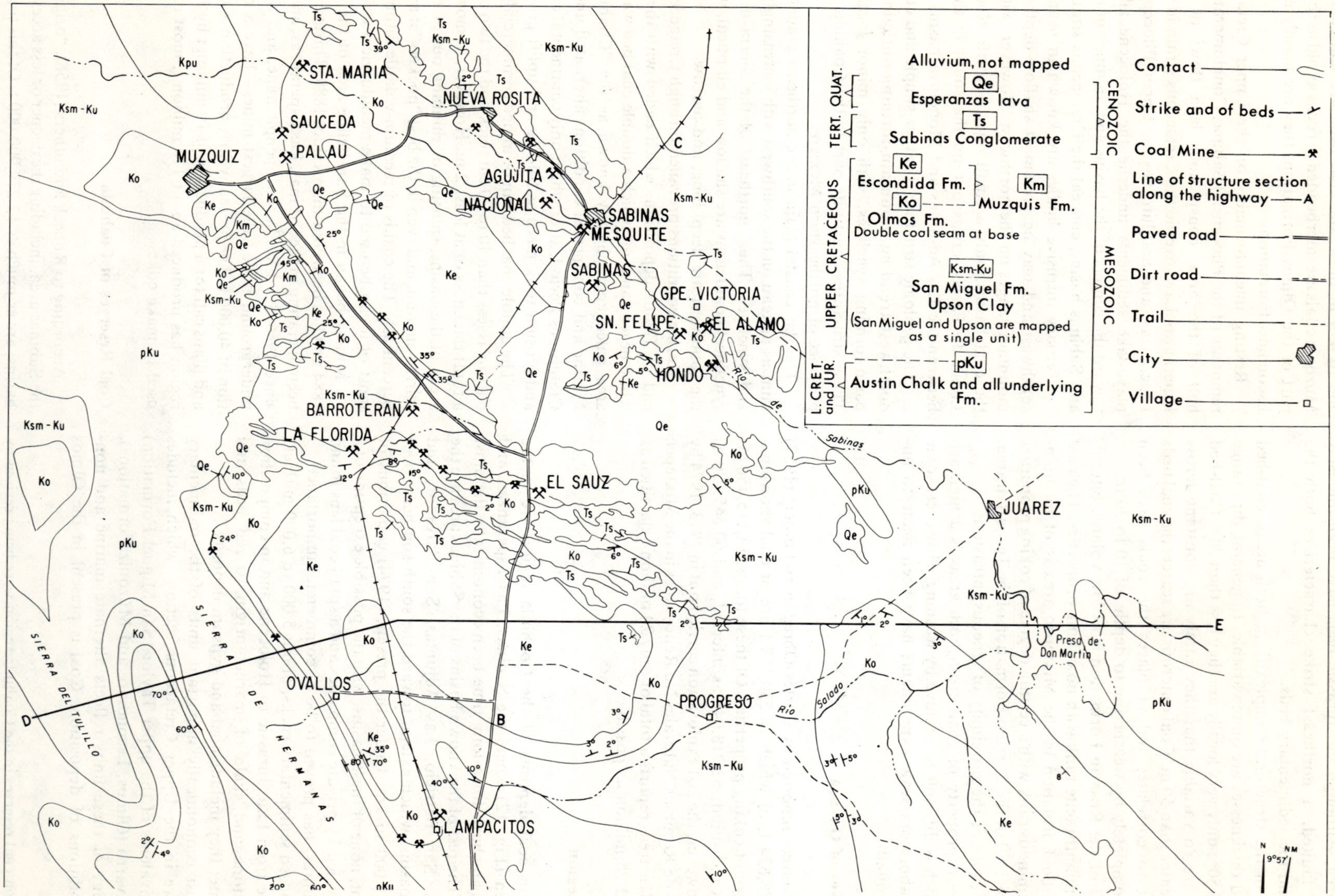

Alluvium, not mapped
Qe
Esperanzas lava
Ts
Sabinas Conglomerate
TERT. QUAT.
CENOZOIC
Ke
Escondida Fm.
Km
Ko
Muzquis Fm.
Olmos Fm.
Double coal seam at base
Ksm-Ku
San Miguel Fm.
Upson Clay
(San Miguel and Upson are mapped as a single unit)
UPPER CRETACEOUS
MESOZOIC
pKu
Austin Chalk and all underlying Fm.
L. CRET. and JUR.
Contact
Strike and of beds
Coal Mine
Line of structure section along the highway
A
Paved road
Dirt road
Trail
City
Village
STA. MARIA
SAUCEDA
PALAU
MUZQUIZ
NUEVA ROSITA
AGUJITA
NACIONAL
SABINAS
MESQUITE
SABINAS
GPE. VICTORIA
SN. FELIPE
EL ALAMO
HONDO
BARROTERAN
LA FLORIDA
EL SAUZ
JUAREZ
OVALLOS
PROGRESO
LAMPACITOS
Presa de Don Martin
Rio de Sabinas
Rio Salado
SIERRA DEL TULILLO
SIERRA DE HERMANAS
C
B
D
E
N
NM
9°57'

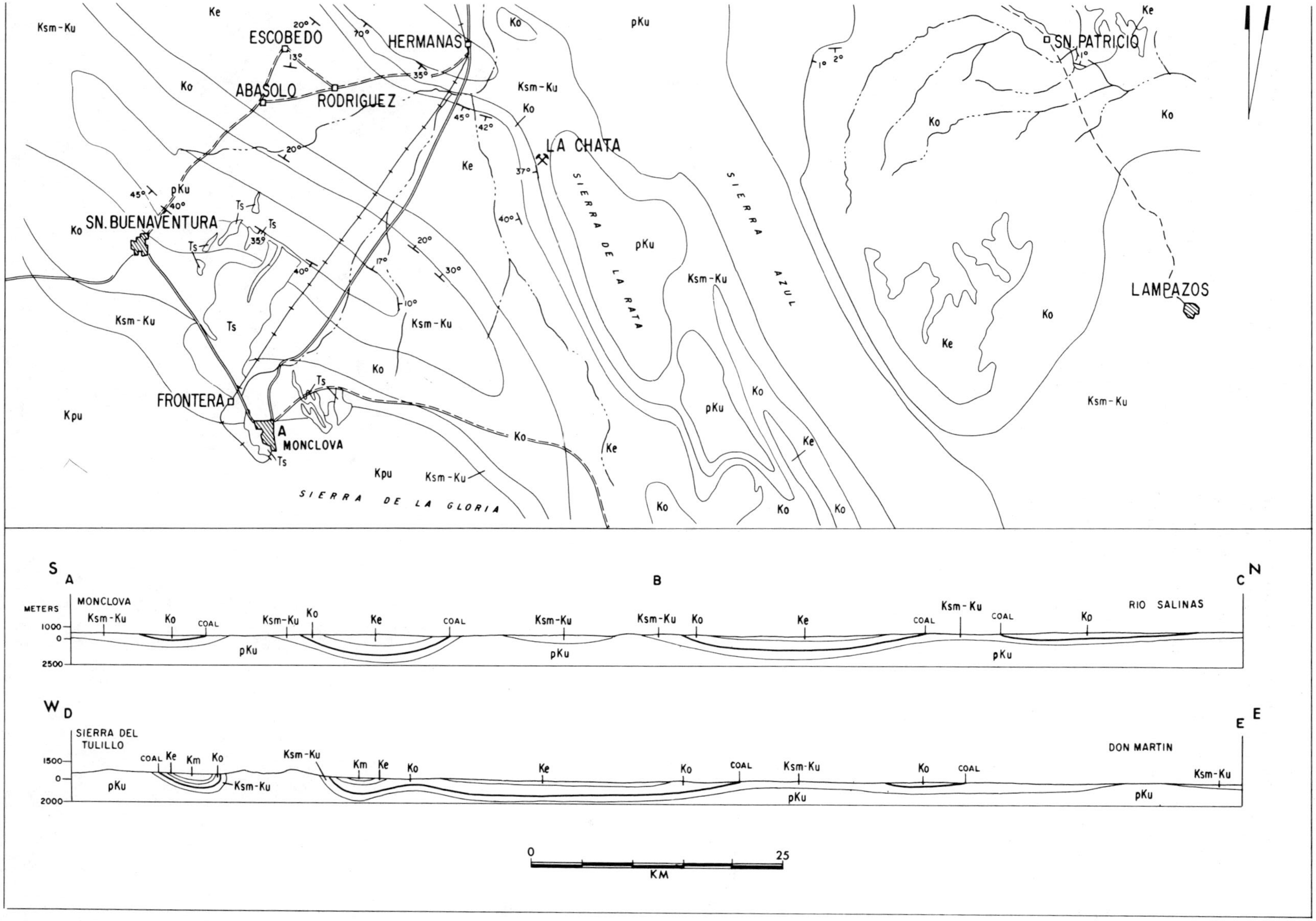

Figure 12. Detailed map of coal-bearing area, Coahuila, Mexico. See Figure 4 for location. From Robeck and others (1956).

reserves and applying several factors, was estimated by those authors to be 1,536,572,000 t.

Through 1972, based on continuous drilling and geologic exploration carried out since 1967 by the Consejo de Recursos Naturales No Renovables in the marginal areas of the Monclova, Las Adjuntas, San Patricio, and Sabinas basins, and on geoeconomic data collected from 1953 through 1972 by the Comisión de Fomento Minero from the operating mines, approximate coal resources of economic importance in the Sabinas region are 243 million t measured and a potential 641 million t recoverable from a total of 1,781 million t estimated in the three most economically important basins (Sabinas, Las Esperanzas, and Saltillito). The rest of the known basins (Las Adjuntas, San Salvador, Lampacitos, San Patricio, and Monclova) contain an estimated reserve of 1,180 million t considered only as a potential resource, most of which is not commercially important under existing conditions of development.

It is evident from a brief comparison between the rate of production and the potential coal distribution that most of the Sabinas basins, mainly the San Patricio basin, still require an extensive, modern program of exploration, based on a sound geologic background.

SELECTED REFERENCES

Camacho, M., Almanza, E., and Juvera, V. M., 1956, Exploración con barrenos de diamante en los depósitos de carbón de Santa Clara, Tonichi, Sonora: Mexico City, Archivo Técnico del Consejo de Recursos Naturales No Renovables (unpub. rept.).

Cortés-Obregón, S., and others, 1957, La cuenca carbonífera de la Mixteca: México City, Banco de México, S.A., Departamento de Investigaciones Industriales, 191 p.

Gonzáles, R. J., 1950, The coal deposits of Coahuila, México: Econ. Geology, v. 45, p. 249–251.

Ojeda R., Jesús, Delgado H., José, and Villamar, S., 1968, Geology of the Sabinas coal basin, Coahuila: Sabinas Coal Region Guidebook, 72 p. (possibly available through interlibrary loan service, U.S. Geological Survey, National Center, Mail Stop 950, Reston, Virginia 22092).

Pesquera V., Rubén, and Carbonell C., M., 1960, Geología y exploración de los depósitos de carbón de la región de San Marcial, Estado de Sonora: Mexico City, Consejo de Recursos Naturales No Renovables, Bol. 59, p. 41.

Robeck, R. C., Pesquera V., Rubén, and Ulloa-Arredondo, S., 1956, Geología y depósitos de carbón de la región de Sabinas, Estado de Coahuila: Internat. Geol. Cong. 20th, 1956 Mexico, D.F., 190 p.

Torón V., Luis, and Cortés-Obregón, S., 1954, Exploration of the Oaxaca coal fields in southern Mexico: Mining Engineering, v. 6, p. 505–509.

Wilson, I. F., and Rocha, V. S., 1949, Coal deposits of the Santa Clara district near Tonichi, Sonora, Mexico: U.S. Geol. Survey Bull. 962-A, 80 p.

——1964, Los yacimientos de carbón de la región de Santa Clara, Municipio de San Javier, Estado de Sonora: Mexico City, Comité Directivo para la Investigación de los Recursos Minerales de México, Bol. 9. 108 p.

MANUSCRIPT RECEIVED BY THE SOCIETY SEPTEMBER 17, 1975
MANUSCRIPT ACCEPTED OCTOBER 17, 1975

Printed in U.S.A.

Geological Society of America
Special Paper 179

Utilization of lignite in the United States

T. Reed Scollon
Eugene T. Sheridan
Mineral and Materials Supply/Demand Analysis, Bureau of Mines, U.S. Department of the Interior, Washington, D.C. 20241

ABSTRACT

The remote location of most United States lignite deposits with respect to major markets and the relatively low calorific value of lignite have prevented, until recently, any significant growth in demand. Unprecedented demands for electric power and recent innovations in its transmission now enable lignite to share with other coals in this growth market. This is particularly true as oil and natural gas resources in the United States decrease and become more difficult to find. Whereas the total reported shipments of lignite had fluctuated between 2.9 and 4.4 million short tons in the decade before 1968, shipments by 1975 had increased significantly to about 19.8 million short tons. The 1975 figure, however, reflects the production of Texas, which was not included in the data for 1956–1968.

In 1975, approximately 4.2 million short tons of U.S. lignite was consumed annually in markets other than electricity—for example, in residential, commercial, and industrial establishments and as a raw material for producing barbeque briquettes, montan wax, and activated char. Although not economically feasible at this time, processes have been developed in which lignite also can be used for the production of chemicals, high-Btu gases, liquid fuels, fertilizers, specialty carbons, and as a reductant for preparation of taconite pellets.

Reserves of lignite are adequate to meet substantial increases in demand, and the costs of recovering the reserves are among the lowest in the United States compared to the costs of recovering reserves of other types of energy. In the immediate future, additional demand for lignite will depend essentially on its increased use to generate electric power. In the long range, nuclear power may limit increases in demand for lignite for power generation. Growth in the utilization of lignite may then depend on the development of the new markets or new products from lignite, such as synthetic fuels. The present United States oil and natural gas shortage also will increase the demand for lignite.

LIGNITE RESOURCES AND DISTRIBUTION

Lignite is the third largest coal resource of the United States. The remaining identified lignite resources are estimated to total 478 billion short tons (1 short ton = 2,000 lb), equivalent on a tonnage basis to nearly 30 percent of the estimated total identified coal resources of all ranks. On the basis of equivalent heating values, however, the lignite resources represent about 20 percent of the total coal resources of the United States. Figure 1 shows these lignite resources related, on a short-ton basis, to resources of other ranks of coal. (The data have been updated throughout the paper to 1975.)

Figure 2 shows that lignite is present in three general areas of the United States: the northern Great Plains, the Gulf Coast, and the Pacific Coast. However, 97 percent of the total resource is in the northern Great Plains. The lignite deposits in this area are in western North Dakota, eastern Montana, and northwestern South Dakota. Appreciable tonnages of lignite are present in north-central and northeastern Wyoming. Yet the bulk of the resource is in North Dakota, and most of the remainder is in Montana.

As shown in Figure 3, North Dakota's lignite resources (estimated at 351 billion short tons) account for nearly three-fourths of the total lignite resources of the United States. Of the Gulf Coast deposits, estimated to total about 12 billion short tons, about 10 billion tons are in Texas, whereas the resources of the Pacific Coast,

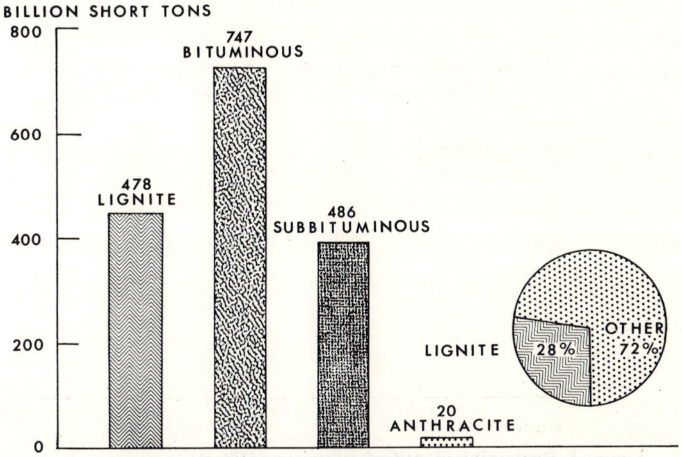

Figure 1. Remaining identified coal resources in the United States on January 1, 1974 (source, *Coal Resources of the United States, January 1, 1974*, U.S. Geological Survey Bull. 1412, 1975).

which are estimated at about 150 million short tons, are principally in the state of Washington.

Estimates for lignite resources are largely inferred. This means that estimates of quantities were based mainly upon an assumed continuity of coal seams between rather widely spaced control points.

Data on the thickness and depth of lignite seams likewise are largely inferred. It is estimated that perhaps two-thirds of the resources are in beds 0.75 and 1.5 m thick. Nearly all of the beds are less than 300 m below the surface. An estimated 80 billion short tons are in beds that range in thickness from 1.5 to 3.0 m thick; at least 247 billion short tons are at a depth of less than 150 m.

The foregoing information indicates that the United States has a very large total supply of lignite. However, the reserve base of lignite—that is, in-place coal deemed mineable under present conditions of economics and technology—is substantially smaller. The reserve base of lignite on January 1, 1976, was estimated at 34 billion tons. This quantity represented only about 8 percent of the total in-place coal of all rank recoverable at this time under current technologic and economic conditions.

As shown in Figure 4, reported lignite production increased slowly from just under 3 million tons in 1956 to about 4.5 million tons in 1967. Lignite production in Texas was not reported to the Bureau of Mines until 1972 and thus was not included in earlier lignite production totals. Lignite production, in millions of short tons, by states, 1972–1975, is shown in Table 1.

As shown in Figure 5, productivity (excluding Texas mines) rose from about 33 tons per man day in 1956 to 58 tons per man day in 1966. Average productivity (including Texas mines) was estimated to be about 80 tons per man day in 1975.

Figure 6 shows that the average value per ton for lignite f.o.b. mines (excluding Texas) decreased from $2.39 in 1956 to $1.98 in 1966. This trend reversed, and by 1975 the average value (also excluding Texas) was $3.17 per ton of lignite.

Lignite production in 1975 came from 10 mines in North Dakota, 4 mines in Texas, and 2 mines in Montana. All U.S. lignite is currently produced by surface-mining methods. Methods differ somewhat, but most lignite operations employ about the same stripping methods as used by bituminous coal operations. In general, stripping consists of removing the overburden in successive adjacent cuts with large draglines and stripping shovels and placing the removed material in an adjacent cut from which lignite already has been removed. At one mine, a bucket-wheel excavator was used for several years, but because of rock inclusions in the overburden, its use was discontinued. The exposed lignite is loaded into haulers by electric-power shovels. The haulers include older bottom-dump units of approximately 20-yd^3 capacity (15 m^3) and end-dump truck units with capacities up to 35 yd^3 (26 m^3). The equipment used for stripping depends chiefly upon the thickness of the overburden. For overburden as much as about 90 ft (27 m) thick, shovels are preferred. Where the overburden is thicker, large draglines are being used in increasing numbers.

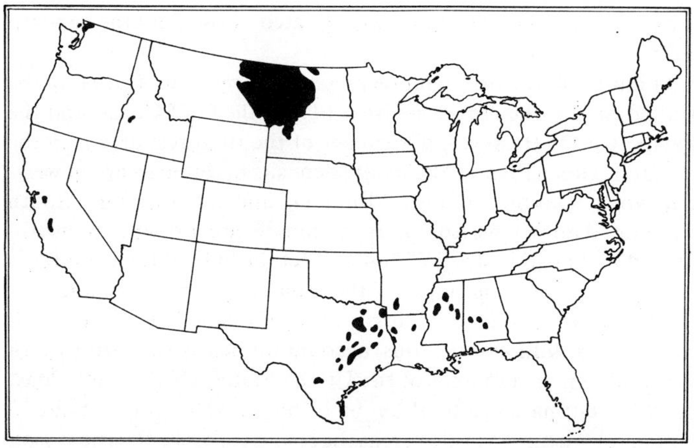

Figure 2. Location of lignite reserves in the United States.

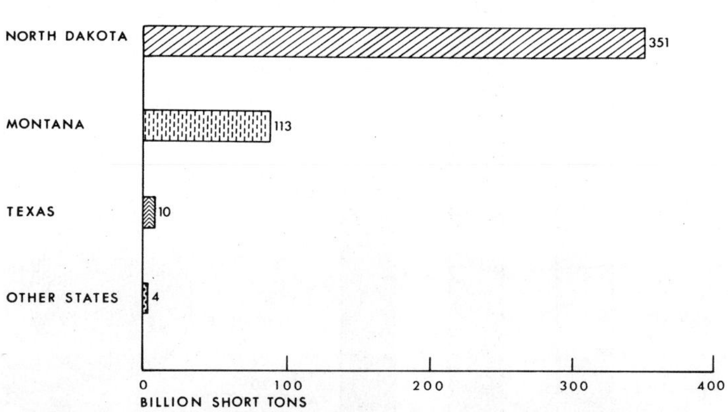

Figure 3. Remaining identified resources of lignite in various states on January 1, 1974; "other states" category includes Alabama, Alaska, Arkansas, California, Colorado, Idaho, Kansas, Louisiana, Mississippi, Nevada, Oklahoma, South Dakota, Washington, and Wyoming.

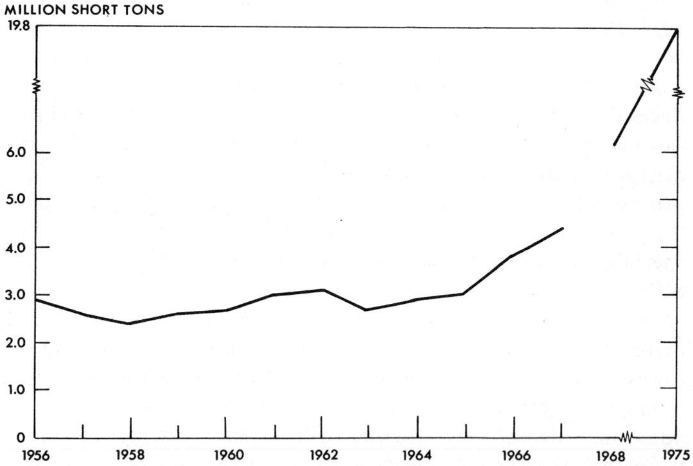

Figure 4. Lignite production in the United States, 1956–1967 and 1975 (source, U.S. Bur. Mines, Minerals Yearbooks).

TABLE 1. LIGNITE PRODUCTION

	1972	1973	1974	1975
Montana	0.3	0.3	0.3	0.3
North Dakota	6.6	6.9	7.5	8.5
Texas	4.0	6.9	7.7	11.0
Total	10.9	14.1	15.5	19.8

Note: In millions of short tons.

LIGNITE USES AND PRODUCTS

Electric Power

Figure 7 records the shipments of lignite by use category in the United States for the period 1956–1967 and 1975. Total shipments of lignite had fluctuated between 2.9 and 4.4 million short tons in the decade prior to 1968, but by 1975 had increased significantly to 19.8 million short tons. The major consumers are the electric utilities, which now account for about 90 percent of the total consumption. Industrial use now accounts for most of the remainder. The data plotted prior to 1967 do not include the lignite produced in Texas.

The use of lignite for generating electric power has increased steadily in the United States because of the growing demand for energy. The construction of large distribution grids and the ability to transmit high-voltage electric power across increasing distances have contributed to the increase.

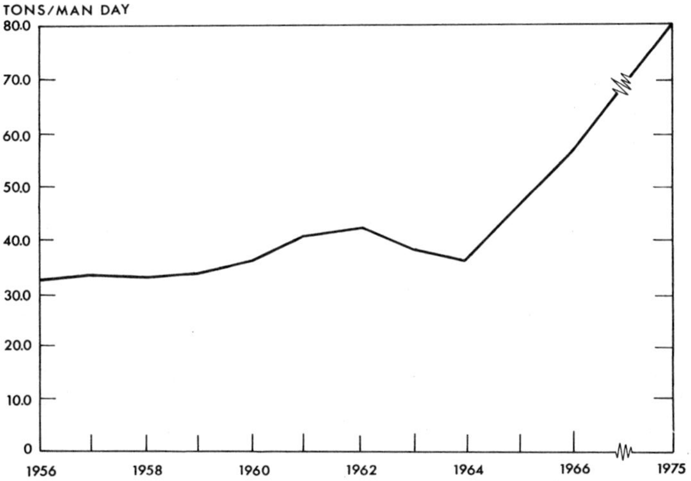

Figure 5. Productivity at lignite mines in the United States, 1956–1966 and 1975 (source, U.S. Bur. Mines, Minerals Yearbooks).

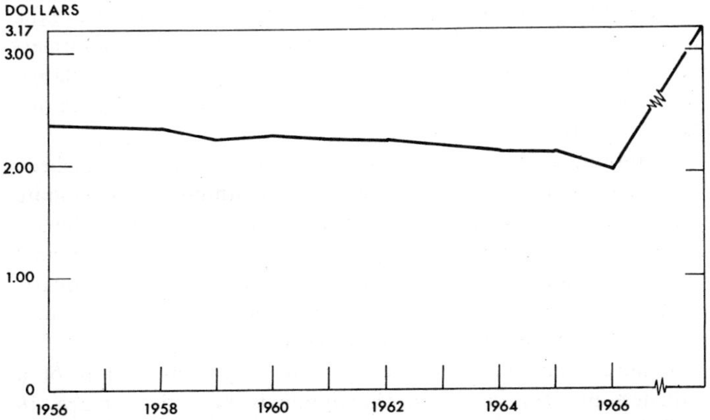

Figure 6. Average value per ton, f.o.b. mines, of lignite produced in the United States, 1956–1967 and 1975 (source, U.S. Bur. Mines, Minerals Yearbooks).

Lignite-Fired Powerplants

Generally, lignite-fired powerplants must have a somewhat larger furnace volume than comparable units burning bituminous coal. The increased combustion space is required because of the modification in flame characteristics caused by the high moisture content of lignite. In many cases, furnaces must be equipped with a greater number of soot blowers to keep conduction surfaces clean. Although combustion efficiencies using lignite are somewhat lower because of the high moisture content, the low fuel cost to plants nearby more than offsets the adverse factors.

For decades, lignite has been burned in powerplants to produce electricity. Until the 1950s, most of these plants were of comparatively small capacity. Some boilers were equipped with underfeed stokers, but the majority were the forced-draft, traveling-grate, spreader-stoker type, which is entirely satisfactory. As the demand for electric power increased, larger units beyond the capabilities of the stoker types were required, and the more recent lignite-fired plants utilize pulverized-fuel firing systems.

A pulverized-fuel system consists basically of a pulverizer to convert the raw fuel to powder, a fan to convey powdered fuel to the furnace, and a burner to mix air and fuel in a manner that will sustain stable ignition. Advantages of pulverized-fuel firing units are rapid response to load changes, wide range of kinds and quality of fuel that may be burned, absence of moving parts in high-temperature sections, and ease of conversion to gas or oil firing. The main disadvantage is the high auxiliary power required for pulverizing.

Another development in coal-burning equipment is the cyclone burner. Designed to burn higher-rank, high-ash coal with low ash-fusion temperatures, cyclone burners can be adapted for lignite firing if some predrying is used to increase flame temperature. In the cyclone furnace, crushed solid fuels and primary air are

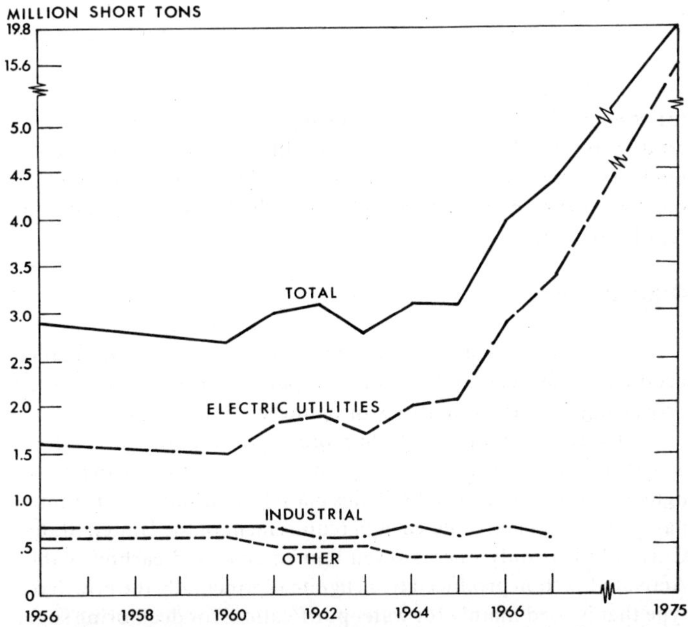

Figure 7. Lignite shipments in the United States, 1956–1967 and 1975 (source, U.S. Bur. Mines, Minerals Yearbooks).

introduced tangentially at one end of a water-cooled, cylindrical, refractory-lined combustion chamber where the fuel is burned at high temperatures as it is moved in a helix pattern by injected secondary and tertiary streams of air. The fuel is completely burned as it moves through the combustion chamber, and molten ash temporarily adheres to the cylinder wall before it flows to a slag tank. This type of burning reduces substantially the amount of ash carried in suspension, and fly-ash emission is low. Advantages of cyclone over pulverized firing are lower auxiliary power requirements (because coal is crushed only to about 6-mm size) and higher efficiency (because of the low excess air and combustible loss in the ash). The main disadvantage is the necessity for a more complicated slag- and ash-handling system.

Lignite-fired electric plants consumed approximately 15.6 million short tons of lignite in 1975. One plant in Texas, the largest lignite-fueled plant in the Western Hemisphere, with an installed generating capacity of 1,187 MW, consumed about 5.4 million short tons of lignite in 1975.

The other principal markets for lignite in the United States are industrial plants that burn lignite as a direct fuel, and retail coal dealers that distribute lignite for use in heating commercial and residential buildings. Lignite in these other markets generally is burned with its as-mined moisture content, which ranges from 35 to 40 percent, in a variety of combustion equipment ranging from hand-fired furnaces and stoves to the more complex furnaces and boilers equipped with traveling-grate or spreader stokers. Markets have been limited generally to areas that are relatively close to producing points. Shipments to these markets in 1975 were about 4.2 million short tons.

Carbonization

Like bituminous coal, lignite can be carbonized and converted into char and chemical raw materials or synthesized into aromatic chemical compounds or liquid and gaseous fuels. However, such processing generally is uneconomical because of the remoteness of lignite-producing areas from markets. Yet for many years, one plant in North Dakota carbonized lignite in Lurgi-Spulgas retorts and converted the char into fuel briquettes, which were used locally for residential and commercial heating. This plant also recovered crude tars, which were processed to obtain creosote-type oil and pitch. The creosote was sold for use as a wood preservative, and the pitch was supplemented by petroleum pitch and used as a binder for fuel briquettes.

Other Products

Lignite also is processed into several nonfuel commercial products. However, only a few companies are engaged in such processing, and the quantities of lignite used are rather small.

Activated Carbon. One of the more important nonfuel products is activated carbon, produced by one plant in Texas using Texas lignite. In operation since 1922, this plant has supplied a substantial part of the total activated carbon requirement of the United States. Currently it is only one of seven known activated-carbon plants. Activated carbon produced from lignite is principally the powdered type that is used mainly for water purification, for decoloring sugar in sugar processing, and for decolorizing dry cleaning solutions. The capacity of the Texas plant is not available, but published data have revealed that its output, combined with plants of another company, accounted for 75 percent of the 190 million pounds of activated carbon produced in the United States during 1966.

Soil Conditioner and Fertilizer. Soil conditioners and fertilizers also can be produced from lignite. Naturally oxidized lignites such as leonardite contain large percentages of humic acids, which benefit soils by improving their physical structure and delaying the release of nitrogen. However, such lignites have a rather low nitrogen content and are not used as fertilizers because they have little nutritive value. Humic acids can be extracted from leonardite and used as the base material for producing a highly nutritive chemical fertilizer with physical properties similar to those of natural soil humus.

Well-Drilling Fluids. Oxidized lignites also are produced commercially for use as viscosity-control agents and gel reducers in well-drilling fluids. Drilling fluids normally have a high content of clay material such as bentonite, which increases the viscosity of the fluid and furnishes a lubricating action and wall-building characteristics. However, drilling fluids that contain bentonite have a tendency to "gel" or to thicken and retain in suspension rock chips, cuttings, and sand that should settle out in mud pits. This condition is eliminated or reduced by the addition of leonardite, which when used in conjunction with an alkali, forms a water-soluble salt that modifies gelation. Leonardite for this use is prepared in two forms: (1) as a fine powder that is mixed with caustic soda at the time of use and (2) as a premixed water-soluble salt that needs no additional alkali. No data are available on the size of this market for lignite.

Montan Wax. One product that may be extracted from some lignite and higher-rank coal is montan wax. Consisting essentially of waxy constituents, such as monohydric alcoholic esters and high-molecular weight organic acids, with smaller amounts of resinous and asphaltic materials, crude montan wax has a high melting point and is hard and brittle. The refined grades, which are compatible with paraffin waxes, are used in a wide variety of commercial products, such as polishes, electrical insulating materials, leather dressings, inks, carbon papers, protective coatings, and greases. Montan wax usually is recovered through solvent extraction with benzene or a benzene-alcohol mixture, and the proportion of the various constituents in the crude product is dependent upon the composition of the lignite and the solvents used in the extracting process.

Commercial production of montan wax in the United States has been limited to Arkansas and California lignites, which are the only known lignites that contain sufficient wax to make commercial recovery feasible. Current production figures are not available, but the total United States output probably now comes from one plant in California.

Uranium. One of the lesser-known products that has been recovered in lignite processing is uranium. Uranium exists in lignite as an oxide, and the uranium oxide content of known uraniferous lignites in the United States is approximately 6 lb (2.7 kg) per short ton. This is from 1 to 2 lb (0.4 to 0.8 kg) more of uranium oxide per short ton than the average amount contained in the other uranium-bearing ores currently being processed in the United States.

Uranium has been recovered from lignite by means of a multistep process in which the lignite is burned in place, in windrows in the field, or in rotary kilns to remove the fixed carbon and volatile matter. Burning without heat utilization is a most crucial step in the process, because to keep the uranium from becoming insoluble, the volatile matter must be driven off and the carbon burned at a temperature below that at which the ash fuses. Uranium oxide is first recovered from the ash by acid leaching and then from the leach solution by solvent extractions. Subsequent processing steps yield final concentrates that are essentially indistinguishable from uranium concentrates recovered from other ores.

The cost of recovering uranium from lignites generally is greater than that of recovery from Western United States sandstone-type uranium ores. This is true because (1) it is necessary to dry and burn lignites before leaching, (2) the acid consumption is considerably higher, and (3) the costs of equipment, operations, and miscellaneous chemicals generally are higher than in the treatment of other ores.

Briquettes. For many years one plant in North Dakota carbonized lignite in Lurgi-Spulgas retorts and converted the char into fuel briquettes that were used locally for residential and commercial heating. This plant also recovered crude tars that were processed to obtain creosote oil and pitch. The creosote was sold for use as a wood preservative, and the pitch was used as a binder for the fuel briquettes.

This briquetting plant now manufactures barbeque briquettes by a process in which iron pyrites are removed from the char, which then is briquetted with a starch binder. Crude tar and creosote also are recovered and processed as in the manufacture of fuel briquettes.

Figure 8 is a diagrammatic sketch of the barbeque-briquette process, the essential steps of which are carbonization of the lignite, separation of pyrites from the char, pulverization of the char, mixing of the char with binder, briquetting, and drying. Carbonization takes place in Lurgi carbonizers where lignite is dried in a predrying section by direct contact with recirculated carbonization bases, then carbonized to drive off the tar and volatile products. Char is removed from the bottom of the carbonizing section by a airlock valve system, after which it passes over a bar screen to remove pyrites; then it is crushed in a hammermill to approximately 4-mesh size. After mixing with starch binder and water, char is briquetted and then dried to 4 percent moisture content in a steam dryer. The briquettes are screened and conveyed to cooling bins where they are allowed to cool for 72 hr.

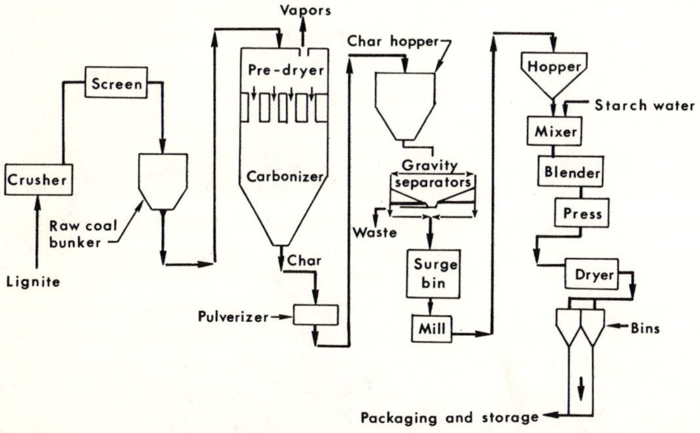

Figure 8. Process diagram for production of barbecue briquettes from lignite.

NEW POTENTIAL LARGE USES FOR LIGNITE

The major uses for and processing of lignite in the United States have been described in the preceding sections. In the section which follows, we discuss some developmental applications that, although not entirely economical, offer potential markets that may increase lignite consumption greatly in the near future.

Gasification. One application that has been investigated intensively by the United States Bureau of Mines and that is the center of renewed interest is the conversion of lignite or other ranks of coal into high-Btu pipeline gas. Gasification is a particularly attractive research area because the development of an economic process for converting coal's energy into gaseous form will enable coal to compete again in heating markets, which currently are dominated by natural gas and fuel oil.

In one basic gasification process, coal is almost completely gasified under pressure with steam and oxygen at 2,500 to 3,000° F to form a mixture of hydrogen and carbon oxides. After purification, the gas is converted at high pressure to methane in the presence of a catalyst at 700 to 900 °F. Another process forms methane directly by the action of steam and hydrogen on pretreated coal or char. The materials are reacted in a two-zone hydrogasifier at about 1,000-psi pressure at 1,200 to 1,900° F. This process has an advantage over the former in that it requires no oxygen and only a minimum of externally produced hydrogen because of internal hydrogen generation.

A 30-ton/day pilot plant that converts lignite into pipeline-quality gas has been constructed at Rapid City, South Dakota. This plant is based on the CO_2-Acceptor process, and the development program is being carried out by Conoco Coal Development Company under a contract with the Energy Research and Development Administration. The CO_2-Acceptor process is based on the use of dolomitic or high-calcium limestone as a medium for heat transfer and removal of carbon dioxide from product gas. This plant has produced up to 1.37 million SCF/day (SCF = standard cubic feet) of 400-Btu gas, and methanation equipment is now being installed to upgrade the product to pipeline-quality gas.

Liquefaction. Another research area offering large potential markets for coal and lignite is liquefaction. As with gasification, technology exists for converting coals into liquid fuel, but the synthetic products are not yet price-competitive with those of crude petroleum. The problem is being investigated intensively, however, and it is hoped that coal also will soon become an important source of the Nation's liquid-fuel supply.

Solvent Refining. The University of North Dakota, under a contract funded by the Energy Research and Development Administration, is currently engaged in a research project aimed at preparing premium fuels from Northern Great Plains area lignites. This project is concerned essentially with laboratory studies that include spectral analyses, titration techniques, and solubility studies of solvent-refined lignite obtained by hydrogenation experiments performed earlier by the Pittsburgh and Midway Coal Mining Company.

Chemical and Metallurgical Uses. Other potential uses for lignite are in the chemical and iron-ore processing industries. Basic chemicals long have been a by-product when agglomerating bituminous coals are carbonized to produce coke. In recent years, research work to convert lignite into chemicals was aimed at producing a high yield of organic solvents and high-volatility liquid-fuel products by the reaction of the lignite with a high-temperature gas.

In iron-ore processing, potential applications for lignite include binding of taconite ores in pelletizing processes and its use as a reductant, both in the conversion of hematite to magnetite before beneficiation and in the manufacture of prereduced iron-ore pellets. The technology for manufacturing prereduced pellets using lignite has been thoroughly demonstrated, but these pellets have been used only in test quantities, and the economics for producing iron by this method have yet to be established. However, competitive pressures may force the steel industry into using prereduced, prepared ore materials. Lignite would be the logical reductant because of the favorable location of lignite resources in relation to nearby iron-ore supplies. Preliminary studies suggest that about 250 lb (114 kg) of fixed carbon, or about 0.5 short tons of raw lignite, is required per short ton of prereduced pellet production. Thus, a pellet plant producing 1 million short tons per year would require about 0.5 million short tons of raw lignite.

SUMMARY

Approximately 3 percent of the total coal used in the United States is lignite. Lignite use has been limited by the remote location of the deposits with respect to major energy-demand areas where higher-rank coal and other fuels are available. Nevertheless, as a result of increasing demands for electric power and developments in high-voltage power-transmission methods, demand for lignite has increased significantly during recent years. The upward trend in this market is expected to continue in the immediate future.

A small portion of the total national output of lignite is consumed in residential, commercial, and other industrial markets. Relatively small amounts are used for nonfuel purposes. Potentials are encouraging for the future use of lignite as a raw material for the production of chemicals, high-Btu gas, and liquid fuels.

MANUSCRIPT RECEIVED BY THE SOCIETY MARCH 30, 1977
MANUSCRIPT ACCEPTED MAY 13, 1977

Printed in U.S.A.